# BASIC STATISTICS FOR THE BEHAVIORAL AND SOCIAL SCIENCES USING *R*

# BASIC STATISTICS FOR THE BEHAVIORAL AND SOCIAL SCIENCES USING *R*

Wendy Zeitlin

and

Charles Auerbach

OXFORD
UNIVERSITY PRESS

OXFORD
UNIVERSITY PRESS

Oxford University Press is a department of the University of Oxford. It furthers
the University's objective of excellence in research, scholarship, and education
by publishing worldwide. Oxford is a registered trade mark of Oxford University
Press in the UK and certain other countries.

Published in the United States of America by Oxford University Press
198 Madison Avenue, New York, NY 10016, United States of America.

Library of Congress Cataloging-in-Publication Data
Names: Zeitlin, Wendy, author. | Auerbach, Charles, author.
Title: Basic statistics for the behavioral and social sciences using R / Wendy Zeitlin, Charles Auerbach.
Description: New York : Oxford University Press, [2019] | Includes bibliographical references and index.
Identifiers: LCCN 2018039059 (print) | LCCN 2018054855 (ebook) | ISBN 9780190620196 (updf) |
ISBN 9780190620202 (epub) | ISBN 9780190620189 (pbk. : alk. paper)
Subjects: LCSH: Social sciences—Statistics.
Classification: LCC HA29 (ebook) | LCC HA29 .Z445 2019 (print) | DDC 519.5—dc23
LC record available at https://lccn.loc.gov/2018039059

9 8 7 6 5 4 3 2 1

Printed by WebCom Inc., Canada

# CONTENTS

# BASIC STATISTICS FOR THE BEHAVIORAL AND SOCIAL SCIENCES USING *R*

# /// 1 /// INTRODUCTION TO BASIC STATISTICS FOR THE BEHAVIORAL AND SOCIAL SCIENCES USING *R*

## INTRODUCTION

Almost all academic programs in the behavioral and social sciences require a course in basic statistics. Some students will ultimately go on to complete their own research and need to conduct statistical analysis independently, while others will primarily become consumers of research. Regardless of how you ultimately use statistics, some foundational knowledge is necessary.

Despite what you may think, you probably already possess some knowledge about the area of mathematics known as statistics. For example, Figure 1.1 displays a pie chart depicting the marital status of 159 people living in an apartment building in New York City.

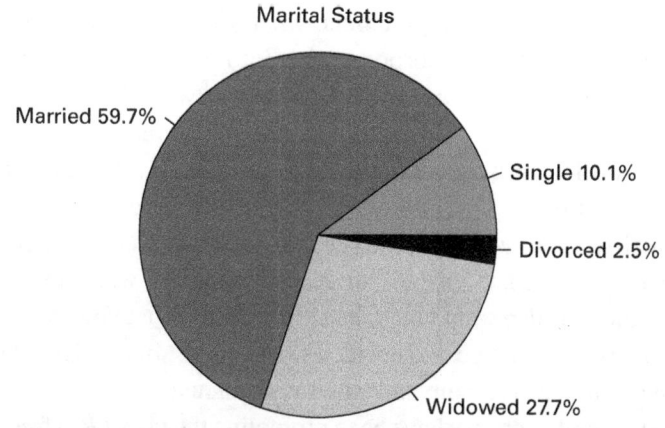

FIGURE 1.1. Marital status of apartment building residents.

Here, you can see that the largest group of people living in the building and who were asked about their marital status is married (59.7%), and the smallest group is divorced, at only 2.5%. The total people questioned (or $N$, as it is called in statistics) is 159 since that is the number of people living in the building who were asked about their marital status. And voila, you already know something about statistics!

That being said, this book doesn't end here and you have not completed your statistics course, so, as we progress through this book, we will try our best to build on things you already know until you have a good understanding of basic statistical principles.

## WHY THIS BOOK IS NECESSARY

If you look at any major bookseller and do a search for "basic statistics" or "introductory statistics," you will literally find thousands of books. So why in the world did we feel the need to write yet another book on this topic?

The answer is pretty simple—we have been teaching statistics in one form or another to social work students for decades. This gives us a pretty good idea of what works well and what doesn't. While we have used some good texts over the years, we wanted to write a textbook for students taking an introductory course in statistics that uses the best of the tools we use when we teach. This, of course, includes covering the standard content that is taught in most introductory statistics courses with good reason; students need to understand, for example, what goes into calculating a standard deviation early on so they can understand more complex concepts, such as standard errors, later. To that end, we think it is important for students to learn how to manually calculate certain statistics and understand their applications; however, we also recognize that after an introductory statistics course, our students will never do this type of manual calculation again.

This is where we start deviating from most other texts. First, we recognize that sometimes a textbook, no matter how good, doesn't reach all students. In our teaching, we sometimes supplement what we teach with videos or other materials, and we decided to incorporate this into our text. In this book, we will frequently refer you to KHAN ACADEMY® (https://www.khanacademy.org/) for specific videos and/or exercises that, we believe, will enhance the content we provide here. Your professor may make use of other KHAN ACADEMY® resources throughout your course.

We also feel strongly about the power that statistics has to increase and share knowledge, so we wanted to enable our readers to apply what they learn here as simply as possible. We also want you to be able to apply what you have learned again and again. To help you out here, we decided to hand you some FREE statistical software that you can use as a student and when you graduate.

For the past several years, we have been promoting the use of $R$, a free open-source statistical programming language that can be used in place of expensive proprietary statistical software that either expires after a short time or can cost thousands of dollars. The reason we have become such big proponents of this is that we have noticed that the organizations with which we work frequently do not have the funds to purchase expensive statistical software. Therefore, our students were not using the statistical packages taught to them once they graduated. We decided that, to build research capacity in not-for-profit organizations, we would be better off teaching our students to use a tool that anyone with a one-time Internet connection could access. Thus began our foray into using R with our students.

Throughout this text, we will replicate what we show you how to calculate manually with R. This will provide you with several distinct advantages. As we said earlier, once you learn how to hand-calculate statistics, you will likely not do it again; therefore, it is helpful from an applied perspective to learn how to produce and read statistical output. With R, this is relatively easy, and you will learn the syntax to produce basic statistical output, often with just a few words, a comma or two, and some parentheses. Not only will we show you how to do this, but also we will provide you with scripts of prewritten R commands that you can modify to conduct your own statistical analysis in the future. In this book, we will show you how to use R, and we will also give you the opportunity to try to conduct your own analysis with a couple of datasets that you will use throughout this book.

## WHY STATISTICS AT ALL?

A world without statistics would, indeed, be a complicated one. Here's a good example. Suppose you are working on a presidential election and want to predict who is going to win. Without statistics, the only way to determine who was going to win in advance would be to ask everyone for whom they were going to vote. This is impractical in a number of ways. First, it would take an inordinately long time to do this—much longer than it would take to actually hold an election. To do this, then, you would have to start polling people far before the actual election, and we know that people change their minds over time. Therefore, not only would this be extremely time consuming and expensive (after all, you actually have to hire a lot of pollsters), but also your results would likely be inaccurate.

A far better idea, then, would be to take a **sample** of the **population** of voters, survey them, and then make inferences about the entire population based on your findings. If you could select a sample (i.e., a portion of the population that you are studying) that has the same characteristics as your voting population, using statistics, you could draw conclusions about all voters. As you can imagine, this is much more efficient in terms of time

For more information on when you would use statistics, watch the KHAN ACADEMY® video "Statistical and non-statistical questions"

and money. Done correctly, statistical inference is amazingly accurate. This is the field of statistics in a nutshell.

In statistics, when you see equations referring to a population, Greek letters will be used. Values associated with a population are called **parameters**. For example, when we were discussing the residents of the apartment building in New York, if we were referring to the marital status of ALL the residents, that would be a parameter. However, if we were referring to the marital status of a sample of residents, which we were, those values are called **statistics**.

When you see equations referring to a sample, English letters will be used. For example, the population mean, or average, is denoted by the Greek letter mu ($\mu$), while the sample mean, or average, is denoted by $\bar{x}$.

## DESCRIPTIVE AND INFERENTIAL STATISTICS

In this book we will use both descriptive and inferential statistics. Descriptive statistics are, not surprisingly, used to describe something. For example, if we wanted to describe the students in your class, we might ask about a number of things:

- Age
- Gender
- What city they come from
- What languages they speak
- What their major is

We could describe the class in terms of the proportion or percentage of people who are male and female. We could also state the average age of the students. Descriptive statistics are basic, but important when you want to boil a lot of information down into something understandable.

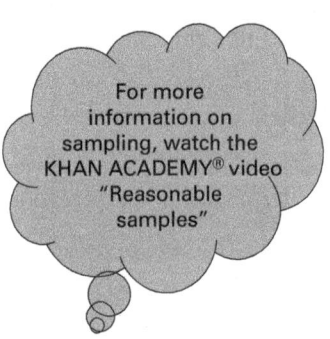

For more information on sampling, watch the KHAN ACADEMY® video "Reasonable samples"

Inferential statistics are more complex. In this area of study, we try to draw conclusions about a larger population based on a sample. To do this well, however, you have to be careful about how you obtain your sample. Your sample should represent the characteristics of the overall population, and the best way to do this, then, is by randomly selecting a portion of the population. This random selection process is designed to reduce **sampling bias**, or the selection of individuals who are different from the larger population in one or more ways based on the methods used to sample. While we do not cover sampling techniques in this book, it is a very important topic in designing research. Please refer to the resources covering research methods, located in Appendix A, for more on this topic.

## LET'S START AT THE VERY BEGINNING...

When we start thinking about statistics, we are thinking about boiling something (or, in most cases, many somethings) down to an idea that is easily comprehensible in some way. Those somethings are called **variables** because they can vary from **observation** to observation. In the behavioral and social sciences, we think of each observation as coming from a person because we are typically interested in people's thoughts, behaviors, and feelings.

Let's think about the example of the residents of the apartment building in New York City. Marital status could be different between residents (and it was, as we saw), so it is a variable. The **indicators**, or possible values attributed to that variable, are the marital statuses that we observed: married, single, divorced, and widowed.

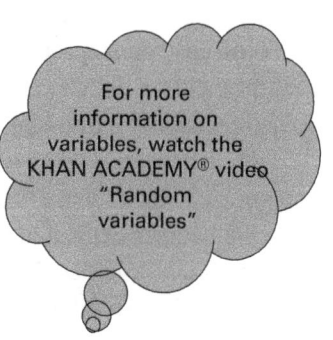

For more information on variables, watch the KHAN ACADEMY® video "Random variables"

In some cases, you will have some things that won't vary from observation to observation. In our apartment scenario, something that does not vary is the fact that everyone lives in a specific building. In this case, where the residents live is not a variable, but rather a **constant** because it doesn't change between observations.

### Levels of Measurement

Variables can be thought of in several ways. First, you can consider the level of measurement of them. You should be concerned about this because level of measurement determines how much precision you get in a variable, and it dictates what sorts of statistical analysis you can conduct later on.

In general, variables can be thought of as categorical or continuous. Categorical, or discrete, variables are simply distinct categories, or named groups, while continuous variables are measured as quantities along some sort of continuum. Categorical variables have less precision than continuous ones.

There are two levels of measurement within the grouping of categorical variables: **nominal** and **ordinal**. Nominal-level variables are categorical variables made up of unranked categories. That is, each indicator cannot be ranked compared to the others. A good example of a nominal-level variable is *marital status*, as discussed earlier, operationalized with the indicators of *married, divorced, single,* or *widowed*. Notice that these categories are discrete and one category does not denote more or less marital status than the other. Variables dichotomized as *yes/no* conditions can also be thought of as nominal. An example of this would be a variable measuring whether someone had a college education. A variable called *college* could be operationalized as *yes* or *no*.

Ordinal-level variables are categorical variables made up of ranked categories. Each indicator can be ranked as greater than or less than in some way as compared to others. An example of an ordinal-level variable could be *level of education* operationalized with the indicators of *less than high school, high school/GED, some college, BA/BS, some graduate education*, or *graduate degree*. Notice that someone who would indicate he or she had *some college* would have less education than someone with a *BA/BS*. In fact, if these indicators were listed on a survey, it would only be common sense to list them in the order described. It would be illogical and confusing to list these indicators like this: *high school/ GED, graduate degree, some college, less than high school, BA/BS*, or *some graduate education*.

When summarizing categorical variables, particularly nominal variables, you will typically report proportions or percentages. When you visualize these, you can present these as pie charts or bar graphs. For instance, we illustrated in Figure 1.1 the percentage of individuals in the sample who fell into each marital status category.

Notice that both types of categorical variables were made up only of words, or categories. None of these was defined by numbers. Some variables, however, are best described numerically, and there are two levels of measurement within the construct of continuous variables. In general, continuous variables are more precise measures than categorical variables.

One type of continuous variable is **interval** level. Interval-level measures denote greater than or less than conditions based on the indicator; however, there is no true zero, which means that it is difficult to describe the true magnitude of difference between indicators. An example of this would be an individual's *level of intelligence* as noted by an IQ score. If one person has an IQ of 100, which is considered average, and another has an IQ of 130, we could state, meaningfully, that the second person's IQ is 30 points higher than the first person's, but you would not conclude that the second person was 30% smarter than the first. It also should be noted that no one has an IQ of zero, which would indicate a total lack of intelligence at all.

**Ratio**-level measures are also numeric, but in these cases, zero is meaningful and denotes the absence of something. For instance, if we were going to measure some aspect of school attendance, we could count the number of days a group of children attended school in the course of a month. If, for one child, we measured 20 days and for another we measured 10, the child with 20 days attended school twice as many days as the child with 10. This means that with ratio-level measures, we can understand a magnitude of difference that was not the case with interval-level measures or with either of the categorical measures.

It should be noted that most concepts can be operationalized in several ways, and it is up to the researcher to identify the best way to measure based on the study's needs. Let's consider the example of level of education. One way to measure this would be to ask people if they had a college degree, which could be answered as a *yes*-or-*no* question. We could also measure this using an ordinal-level measure by asking them their highest level of education, as illustrated earlier, and we could provide the following choices: *less than high school, high school/GED, some college, BA/BS, some graduate education,* or *graduate degree.* We could also simply ask people how many years of education they have completed, and a number could be obtained, which would be a ratio-level measure (although this may not actually provide us with the information we would want since some people take more than four years to obtain a bachelor's degree). If we obtained this information as a nominal-level measure, there would be no way to determine how many years of education people had who answered "yes" to whether they had a college degree. Similarly, if we asked this as an ordinal-level measure, we could collapse answers into the nominal-level categories, but we could still not determine how many years of education an individual had. If, however, we were to ask this as a ratio-level measure, we could possibly determine which categories individuals fell into in the ordinal and nominal measures based on who was in our sample. This example is illustrated in Table 1.1. Notice that we did not measure education as an interval because we simply were not able to determine an adequate way to measure the concept in this way. Notice also that, in this case, you are most likely interested in an ordinal-level measure. You should, however, be aware of the level of precision achieved at various levels of measurement.

TABLE 1.1. Measuring *Education* as a Variable With Different Levels of Measurement

| Level of Measurement | Description | Example of Measuring Homelessness | | |
|---|---|---|---|---|
| Nominal | Unranked categories | Degree/no degree | | Less precision |
| Ordinal | Ranked categories | Less than high school, high school/GED, some college, BA/BS, some graduate education, graduate degree | | |
| Interval | Numeric values with no true zero | | | More precision |
| Ratio | Numeric values with a true zero | Actual number of years of education | | |

**Some Additional Considerations Regarding Level of Measurement**

Level of measurement is important to consider for two reasons. First, as discussed previously, you will want to consider the level of precision you need in measuring any one variable. The other is that the way you evaluate your data is largely dictated by the level of measurement of your variables. As you will see in the next chapter, for example, we typically summarize categorical variables in terms of the percentage of the sample that falls into each category, and continuous variables are typically summarized by reporting the mean, or average, value of the variable. For instance, while we could report that 50% of our sample has a bachelor's degree, the continuous variable *age* might be better described by saying something like, "The average age of the people in our sample was 24.3 years."

Because of this, we want to pay special attention to features of some types of variables:

1. In some instances, ordinal variables are best thought of as an interval. Consider the example of a survey in which individuals are asked to assess how much they agree with the following three statements designed to measure students' satisfaction with their professors:

| Statement | Strongly Disagree | Disagree | Agree | Strongly Agree |
|---|---|---|---|---|
| My professor is an effective instructor | | | | |
| I can get in touch with my professor easily outside of class | | | | |
| My professor is an expert in his or her field of study | | | | |

While each of these items is an ordinal variable, if we assigned numbers to each of the categories ranging from 1 = strongly disagree to 4 = strongly agree, we could treat the entire instrument as an ordinal scale and be able to do more extensive analysis. For instance, Student #1 agreed with the first two statements and disagreed with the third. Student #1's total level of professor satisfaction could be calculated by summing each variable for a total score of 8. This score could then be compared to that of other students.

2. Some ratio-level variables are counts. For instance, we could count the number of days that college students missed class or we could count the number of times that a patient went for physical therapy. Count variables sometimes present some additional considerations when we include them in our analyses because their values are typically clustered. We will discuss this further later on.

### Relationships of Variables to One Another

Often, you will be concerned with examining the relationship between one variable and others. In most cases, the desired result is your **dependent variable**, which is sometimes also referred to, not surprisingly, as an outcome variable since it is typically the outcome in which you are most interested. Variables that you think will be predictive of the dependent variable are known as **independent**, predictor, or explanatory **variables**. In this book, we will look at variables separately, but we will also think about the relationship two variables may have to one another.

### Basic Statistics Involves a Little Math

You don't need to be a math whiz to get through an introductory course in statistics, but some basic math and algebra skills are helpful. If you haven't taken a mathematics course recently or feel a little rusty on the following topics, you should refer to Appendix B, which points you to some resources that cover the following topics:

- Basic arithmetic—including order of operations, sigma notation, scientific notation, and cancelling in fractions (i.e., reducing fractions to their lowest terms);
- Basic algebra—this includes an understanding of topics such as understanding and writing simple algebraic expressions, combining like terms, and rearranging expressions to isolate a single variable;
- Linear equations—understanding the equation for a line and graphing; and
- Factorials, Combinations and Permutations—this includes examples illustrating the differences between these and how factorials fit into the equations.

If you feel rusty at all on any of these topics, refer to those resources for more review and practice.

### GETTING STARTED WITH *R*

We like *R* a lot! It is a powerful tool that can be used for statistical analysis, it is open source, and it is FREE! What this means is that people worldwide develop packages, or specialized bits of code, to do thousands of things, including managing data, statistical analysis, and publishing. Ultimately, what this means to you is that by learning to use *R*, you can avail yourself of these for absolutely no money whatsoever.

If you get involved in conducting your own research, *R* is probably the most powerful tool available for making research reproducible. This enables other researchers to re-create your analysis by looking, at the very least, at the code you used to conduct your own analysis (Gandrud, 2015).

As we mentioned previously, once you learn fundamental skills in statistics, it is unlikely that you will choose to manually calculate a statistic ever again. That's the good news.

The more complicated thing, though, for budding statisticians is how to interpret output from statistical programs. To address this, we will illustrate each of the statistical techniques we show you in *R*, and you can download the data used in examples and exercises used in this book from our website, http://www.ssdanalysis.com, or from Oxford University Press's website for this book.

Complete download and setup instructions for *R* can be found in Appendix C. This includes how to download *R* and *RStudio*, a free graphical user interface that makes using *R* simple. Appendix C will also introduce you to *R* packages, which are collections of user-contributed functions that will allow you to do more statistical analysis (and some other things) than you will ever need. Appendix C will also introduce you to some basic *R* functionality such as opening datasets, creating and defining different types of variables and vectors, assigning values to variables and vectors, and saving datasets.

Appendix C is not a comprehensive primer on *R*. Appendix A contains a list of some of our favorite references for more information about using *R*. We wholeheartedly recommend that you not go overboard and try to learn anything and everything you can about *R*. Rather, we recommend that you learn what you need to when you need it. Because of its enthusiastic and generous user base, there are many excellent and free resources available via the Internet along with excellent books.

To help make using this book a little easier, we will list the packages used in each chapter at the beginning of that chapter.

## GETTING STARTED WITH KHAN ACADEMY®

As you may have noticed, we have already introduced you to the most basic way this website can be used with this book. We will suggest KHAN ACADEMY® videos that we think supplement content throughout the book. Whenever you see a thought bubble, like the one illustrated in Figure 1.2, read it carefully for the topic covered in a suggested video along with the title of the video. In the example, we tell you that the topic covered in the video is sampling, and the name of the video is "Reasonable samples."

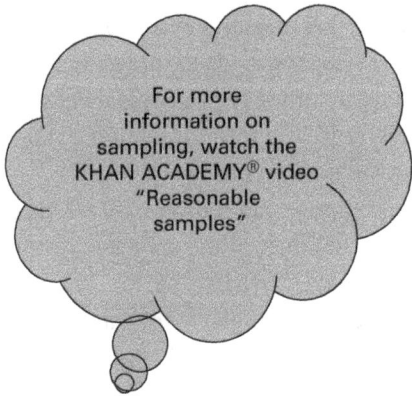

FIGURE 1.2. Example of suggested video.

To access any of the recommended videos, simply go to http://www.khanacademy. org, as illustrated in Figure 1.3.

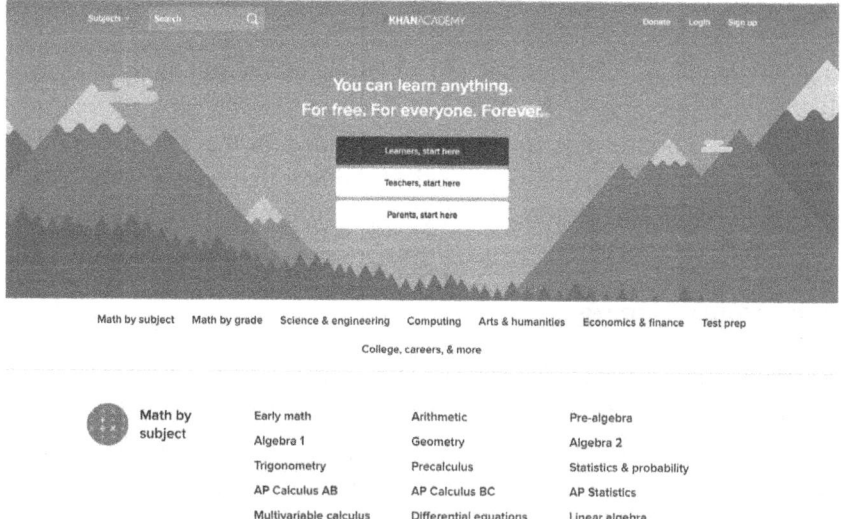

FIGURE 1.3. KHAN ACADEMY® home page.

To access any of the videos or exercises we specifically recommend, simply enter the video name in the search box.

## EXAMPLES

Throughout this book, we will provide you with many examples. You will have the opportunity to observe and follow step-by-step examples demonstrating each statistical procedure. You will also be able to practice each manually, and then try it out using *R*.

We have learned, however, that data does not generally exist in a vacuum; that is, you, as a student, researcher, or statistician, will have some specific interest in your data because it will pertain to some work you are doing. To make this text meaningful, then, we are providing you with three datasets that you will become immersed in throughout the course of this book. The first dataset, called *hospital*, contains information about 50 patients admitted to County General Hospital through its emergency room. The data from *hospital* will primarily be used to demonstrate how to calculate and interpret the various statistics we present in this book in *R*.

The other dataset, called *workers*, contains information about 30 employees at the Funky Fun Corporation. We will only use this dataset in Chapter 7 because of some of its unique qualities.

In addition, we have included a spreadsheet called *students* in Appendix D that you will use for hand-calculating exercises shown at the end of each chapter. These data are based on 20 fictional students attending YOUR school!

Appendix D describes each of these datasets in more detail and presents the data for both *hospital* and *workers* in a table.

## TERMINOLOGY

Throughout this book, we use standard mathematical terminology and symbols. Words that you see written like **this** will be found in the glossary in Appendix E. Definitions and brief explanations will be provided for these terms. Additionally, we have created an appendix of mathematical symbols that appear in the text. These can be found in Appendix F.

## HOW THIS BOOK IS ORGANIZED

This book is divided into two sections. The first is made up of 10 chapters. Together, they form the basis of a first-level course in statistics. We begin with this introductory chapter. We would encourage you at this point to review the basic mathematical concepts covered in Appendix B. If you are familiar with these concepts, you can easily move on. If not, we would encourage you to review these ideas through both resources suggested in Appendix B and videos and activities recommended on the KHAN ACADEMY® website.

Chapter 2 continues the discussion we began in this chapter on variables. Here we talk about the numerous ways in which you can understand and visualize both categorical and continuous variables.

In Chapter 3, we talk about the notion of probability and address examples such as the chances of rolling particular numbers on a die or being dealt combinations of cards. This is foundational to the study of statistics since the majority of this field is built on probabilities given particular circumstances and how we might expect data to be distributed. As we move through the rest of this text, we will frequently refer to concepts introduced in this chapter.

Chapters 4 and 5 extend some of the ideas covered in Chapter 3 and then expand upon them by introducing inferential statistics. Here, we discuss using what we know about a sample and determine the degree to which we can apply it to a population. You will learn about the normal distribution, or what is often referred to as "the bell curve," and you will learn why it's important in statistics. In these chapters, we will also help you identify hypotheses that you might want to test.

Chapters 6 through 9 give you the skills to test hypotheses based on the nature of your data. Chapter 6 shows you how to test hypotheses that include comparing means between samples when your data are normally distributed, and Chapter 7 shows you how to test hypotheses that include comparing means between samples when your data are not normally distributed. Chapter 8 teaches you about correlation and its cousin, linear regression, while Chapter 9 covers tests of association. While these terms may be confusing to you now, they will be cleared up as you move through this text.

Chapter 10 moves away from hypothesis testing and addresses power analysis, which is often helpful in designing research studies and choosing appropriate sample sizes.

The second and final section of the book contains the many appendices described throughout the text. While these appendices do not form the core of this book, they are helpful additions, which we believe you will find useful:

Appendix A—Resources—websites and other books for additional information
Appendix B—Math review
Appendix C—All about $R$
Appendix D—Example datasets
Appendix E—Glossary of statistical terms
Appendix F—Tables of distributions

## EXERCISES

1. Find a current news article that contains a display of data. This could include a pie chart, infographic, or any other representation of quantitative data. Why do you think the author(s) included this? What do you think this article may have been like without this display?
2. Find a current news article that contains statistics of some sort. How do you think the reporting of these statistics enhances or detracts from this article?
3. For each of the following scenarios, what would be the dependent variable and what would be the independent variable? If there is a constant in the scenario, state what that is:
   a. You want to better understand if cognitive behavioral therapy (CBT) has an impact on combat soldiers' posttraumatic stress disorder (PTSD).
   b. You and your friends want to see if there is a relationship between what students major in in college and what their incomes are five years after they graduate.
   c. Your psychology professor tells you that she thinks students who study in the library get better grades than students who study in their rooms. Because you are open to studying in either place, you decide to figure out whether or not this is true.
   d. You have to do a research paper for your sociology course. You are asked to determine whether those who live in urban areas are more or less likely to have racial prejudices compared to those living in suburban or rural areas.
4. For each of the following variables, describe in detail how you would operationalize each as a categorical or continuous variable. Then state whether

categorical variables were operationalized as nominal or ordinal and whether continuous variables were operationalized as interval or ratio:

a. Religious preference
b. Age when you get your first full-time job
c. Household income
d. Sex
e. Satisfaction with your statistics course
f. Major in school
g. Grades
h. Self-esteem

# DESCRIPTIVE STATISTICS

---

To work through the examples in this chapter, you will need to *install* and *load* the following *R* packages:

- *psych*
- *vcd*
- *fdth*
- *Hmisc*

You should also load the following dataset into the Environment pane:

- *hospital*

For more information on how to do this, refer to Appendix C.

---

## INTRODUCTION

In Chapter 1, we introduced the idea of variables to you. In this chapter, we expand on that notion by showing you how to boil down a lot of information about many observations of variables into something that makes sense and describes them in ways that are easily understood by others.

This is done in two ways. First, we will show you how to calculate descriptive parameters or statistics for single variables (hence the term "univariate statistics"). And, because we believe the saying that "a picture is worth a thousand words," we also show you how to do this visually. In calculating descriptive statistics, we show you how to do this manually, but then we show you how to do these same calculations using *R*. When we teach you how to visually display descriptive statistics, however, we only show you how to do this in *R*, as it is highly unlikely that you would ever need to, say, draw a bar graph or histogram by hand. Learning how to do this using software, however, is a powerful way to explain your data.

As you begin to think about describing your data, you will need to consider the level of measurement of your variables. If, for example, you wanted to describe the ages of the students in your class, it would be more difficult to understand the "typ-ical" student by listing how many 18-, 19-, 20-, 21-, and 22-year-olds there are than if you gave the average age of your classmates. Similarly, if you were to look at the marital status of the NYC apartment dwellers described in the previous chapter, you could not tell someone the average marital status of the building's residents, but you might tell him or her the category with the highest number or percentage of people in it. Therefore, as we move through this chapter, we will begin by thinking about levels of measurement from the least precise to the most precise.

## CATEGORICAL DATA

### Proportions and Percentages

A sample of students participating in a psychology research study may consist of, for example, 185 females and 101 males. For comparison reasons, it may be neces-sary to take into account the number of respondents in each group. Both proportions and percentages take the number of observations in each category into account by standardizing the categories by their relative size.

A basic assumption in reporting proportions and percentages is that categories are mutually exclusive (i.e., a subject can be in only one category) and exhaustive (i.e., all possible categories are represented). For example, let's take the nominal variable "marital" in the *hospital* dataset included with this text. *Marital* includes four distinct categories: *divorced, married, single,* and *widowed.* The counts of the individuals in each category are *divorced* = 11, *married* = 23, *single* = 12, and *widowed* = 4, with the total $N = 50$:

$$11 + 23 + 12 + 4 = 50$$

The proportion of people who are divorced is:

$$\frac{\# \, divorced}{N} = \frac{11}{50} = 0.22$$

Similarly, the proportion of people who are married is:

$$\frac{\# \, married}{N} = \frac{23}{50} = 0.46$$

Using this same formula, the proportion of people who are single is 0.24, and the proportion of people who are widowed is 0.08.

When you add all the proportions for each of these categories we get 1, or unity, and you will never be able to exceed 1 for a given variable because that encompasses all people who were studied:

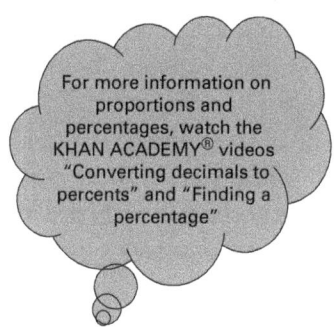

For more information on proportions and percentages, watch the KHAN ACADEMY® videos "Converting decimals to percents" and "Finding a percentage"

$$0.22 + 0.46 + 0.24 + 0.08 = 1.00$$

Percentages can be obtained from proportions by multiplying each proportion by 100. Therefore, the term *percent* means "per one hundred." In the case of a percentage, unity is 100, as 100% of something represents the entirety of that thing. Percentages are used much more often than proportions.

To convert our first calculated proportion of 0.22 to a percentage, we can do the following:

$$0.22 \times 100 = 22\%$$

Similarly, we can convert 0.08 to a percentage:

$$0.08 \times 100 = 8\%$$

## OBTAINING PROPORTIONS AND PERCENTAGES USING *R*

The **describe()** function in *Hmisc* in *R* can be utilized to obtain counts and proportions for a particular variable. To demonstrate this, open the *hospital* dataset as described in Appendix C. Type the following into the Console:

>**describe(hospital$marital)** and press <ENTER>

The results are displayed in Figure 2.1.

```
> describe(hospital$marital)
hospital$marital
        n  missing distinct
       50        0        4

Value      Divorced  Married  Single  Widowed
Frequency        11       23      12        4
Proportion     0.22     0.46    0.24     0.08
```

FIGURE 2.1. Counts and proportions of individuals by marital status.

Notice that this output tells you the total sample size for this variable ($n = 50$), how many missing observations there are (none), and how many categories of *marital* we have (four). Then the counts for each are shown in the row labeled *Frequency*, and the proportions for each are shown directly below.

These results, which were obtained using the same calculations shown in the previous section, explain that almost half of the people in the sample are married while nearly a quarter are single. The smallest group of people, only 8%, were widowed.

### Displaying Nominal and Ordinal Data

Nominal and ordinal data can easily be displayed using bar charts. Figure 2.2 displays a bar chart of the counts of marital status that we obtained previously. To produce this figure, we will first create a vector, which we will call *m1*, to sort all the data by marital status. Then we will use that sorted data to build our bar chart. Notice that we chose to add labels to both the x-axis and y-axis, and we instructed *R* to have the y-axis start at a value of 0 and end with a maximum of 25 (since we know from our calculations earlier that the largest category had 23 observations). Enter the following into the Console:

```
>m1<-table(hospital$marital)
>barplot(m1, ylab="count", xlab="marital status",
ylim=c(0,25))
```

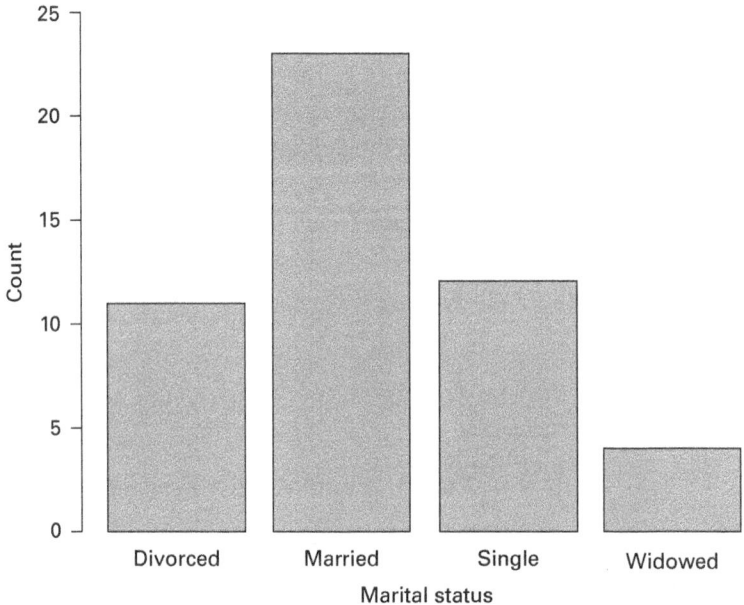

FIGURE 2.2. Bar chart of counts of marital status.

The addition of the **prop.table()** function to the syntax will create a bar chart of the proportions (Figure 2.3). Notice how we changed the label of the y-axis to reflect that we are now showing proportions, and the limit on the y-axis is set at one to reflect unity.

```
>barplot(prop.table(m1),ylab="proportion",xlab="mari
tal status", ylim=c(0,1))
```

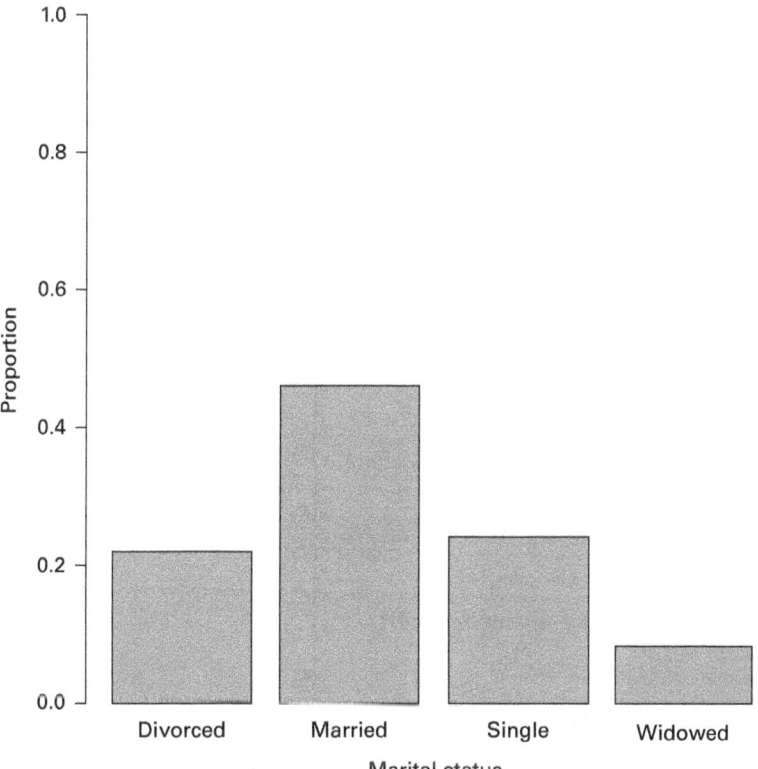

FIGURE 2.3. Graph of proportions of marital status.

This same idea can be extended to compare two variables that are both measured categorically. To illustrate this, consider just 10 patients from the *hospital* dataset. We can think about whether men or women are more likely to be admitted to the hospital from the emergency room. Notice that both *gender* and *EDadmit* are categorical variables. Table 2.1 shows this data.

TABLE 2.1. Data Showing *gender* and *EDadmit*

| gender | EDadmit |
|--------|---------|
| Female | Yes |
| Male | No |
| Male | No |
| Male | No |
| Female | Yes |
| Female | No |
| Male | Yes |
| Female | Yes |
| Male | Yes |
| Male | Yes |

You may want to illustrate whether men or women were more likely to be admitted to the hospital from the emergency department. You can construct a table like the one shown in Table 2.2 simply by counting the number of people in each of four categories from Table 2.1:

- Males admitted from the emergency department
- Males not admitted from the emergency department
- Females admitted from the emergency department
- Females not admitted from the emergency department

TABLE 2.2. Categories of Emergency Department (ED) Admissions by Gender

| ED Admit | Gender | | Totals |
|----------|--------|------|--------|
| | Female | Male | |
| Yes | 3 | 3 | 6 |
| No | 1 | 3 | 4 |
| Total | 4 | 6 | 10 |

From this table, we can see that there was a total of four female patients and six male patients. Six people, three females and three males, were admitted from the emergency department, and four people, one female and three males, were not.

Since we know that there were 10 patients admitted altogether, we can calculate proportions by dividing the count in each cell of the table by the total number of patients. Therefore, the proportion of all patients who were females admitted from the ED is:

$$\frac{3\ females\ admitted\ from\ ED}{10\ patients\ in\ all} = 0.30\ or\ 30\%$$

Similarly, we can calculate the proportion of all patients who were females and not admitted from the ED:

$$\frac{1\ female\ not\ admitted\ from\ ED}{10\ patients\ in\ all} = 0.10\ or\ 10\%$$

Therefore, we could create a similar table to the one previously, but Table 2.3 uses proportions instead of counts.

TABLE 2.3. Proportion of Emergency Department (ED) Admissions by Gender

| ED Admit | Gender | | |
| --- | --- | --- | --- |
|  | Female | Male | Totals |
| Yes | 0.30 | 0.30 | 0.60 |
| No | 0.10 | 0.30 | 0.40 |
| Total | 0.40 | 0.60 | 1.00 |

While this is relatively easy to do by hand for small datasets, doing this with larger datasets is time consuming and could lead to errors in calculations. Therefore, it is helpful to learn how to do this same thing in *R*.

To illustrate this, we can depict admissions to the hospital from the emergency room by gender for all 50 patients in the *hospital* dataset. Now we can construct two-way tables, like those shown previously, describing counts and proportions for each group, much as we did manually:

```
>admit<-table(hospital$edadmit,hospital$gender)
>admit
```

The results you get here are displayed in Figure 2.4.

```
      Female Male
No         7   17
Yes       16   10
```

FIGURE 2.4. Patients sorted by gender and admission from emergency department.

We can gather even more useful information if we look at the proportion of males and females admitted to the hospital through the emergency department, and we are going to add an option to total these proportions by column. This will tell us what proportion of all females were admitted through the emergency department along with what proportion of all males were admitted through the emergency department. Enter the following into the Console:

```
>prop.table(admit,2)
```

The results are displayed in Figure 2.5.

```
> prop.table(admit,2)

          Female      Male
 No    0.3043478 0.6296296
 Yes   0.6956522 0.3703704
```

FIGURE 2.5. Table of proportions of emergency department admissions by gender.

The vector *admit* is used in the **prop.table()** function because it contains our two sorted variables. Notice that in the **prop.table()** function, the integer "2," which follows a "," is an option included to instruct *R* to produce column proportions. The results suggest that, of all the females, nearly 70% are admitted through the emergency department, compared to only 37% of the males. Also notice that if you add the column proportions for females only, you will get 1, indicating that we are looking at the proportions by column title (i.e., female and male).

It can be misleading to provide only percentages when the number of cases is small, but many of us are accustomed to considering only percentages and ignoring the counts of observations. In the previous example, when we only considered proportions, it seemed very clear that women are more likely to be admitted through the emergency department than men; however, when we looked at the counts of each cell, we saw that this sample was so small that it would be hard to make any general statements about anyone at all. As a rule, then, the number of observations should be reported along with percentages.

As displayed in Figure 2.6, the two-way table in Figure 2.4 can also be displayed as a bar plot using the syntax that follows. Most of the time you would want to display your findings in a graph as a percentage, but, as discussed earlier, when the number of cases (the *n*) is small, percentages can be misleading. As a result, we have shown how to produce a graph with counts here:

```
> barplot(admit, xlab="gender", ylab="count",
main="Gender of Patients by Admission from Emergency
Department", legend=c("No", "Yes"), ylim=c(0,30))
```

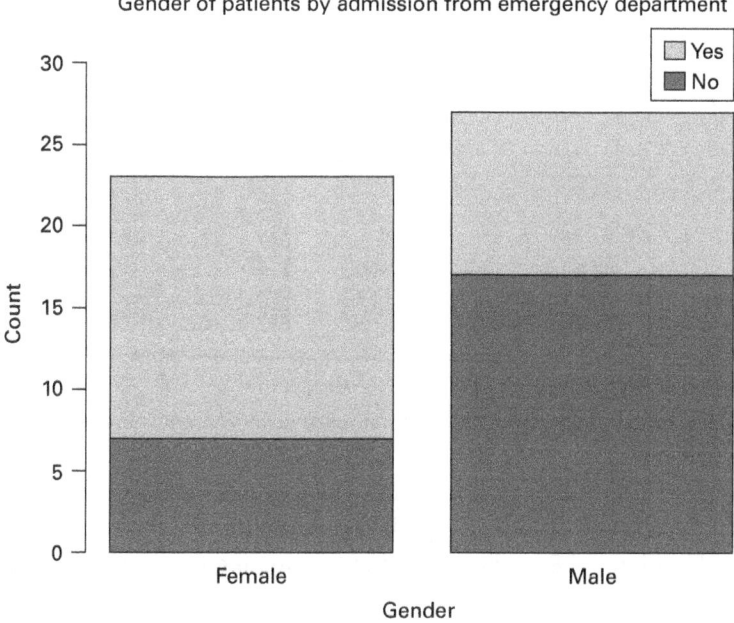

FIGURE 2.6. Admissions from emergency department by gender.

In the previous syntax, notice that we added a legend to the bar plot to show that the light gray boxes denote "Yes" and the dark gray mean "No." To display the legend accurately, you will want to place the levels of the first variable in the table you created in alphabetical order; this is why, in the **legend** option, "No" was listed before "Yes." We also enhanced this graph to add labels to both the x- and y-axes along with a main label. Notice that all labels are put inside parentheses.

To display proportions instead of counts, as illustrated in Figure 2.7, you only need to modify the previous function slightly. To do this, enter the following into the Console:

```
> barplot(prop.table(admit), xlab="gender",
ylab="proportion", main="Gender of Patients by
Admission from Emergency Department", legend=c("No",
"Yes"), ylim=c(0,1))
```

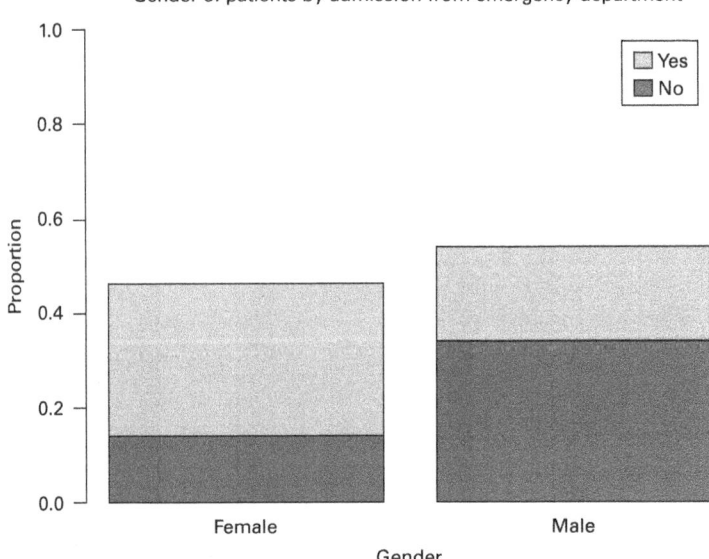

FIGURE 2.7. Proportion of admissions from emergency department by gender.

Finally, you could display these same data as percentages by multiplying the proportions by 100 and adjusting your labels to accurately reflect this. This bar plot is illustrated in Figure 2.8, and the function for creating this is as follows:

```
> barplot(prop.table(admit)*100, xlab="gender",
ylab="%", main="Gender of Patients by Admission
from Emergency Department", legend=c("No", "Yes"),
ylim=c(0,100))
```

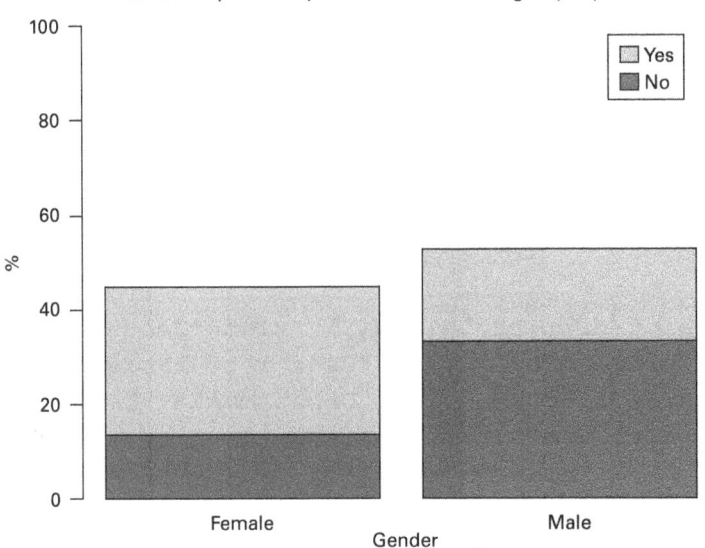

FIGURE 2.8. Percentage of admissions from the emergency department by gender.

In the previous examples, we showed you how to create stacked bar plots, but you may want to present this data slightly differently: by showing the bars side by side. Doing this is easy: You will simply add one more option, **beside = T**, to the **barplot()** function. To try this, enter the following in the Console, and you will see the results displayed in Figure 2.9 in the Plots pane:

```
> barplot(prop.table(admit)*100, xlab="gender",
ylab="%", main="Gender of Patients by Admission
from Emergency Department", legend=c("No", "Yes"),
ylim=c(0,100), beside=T)
```

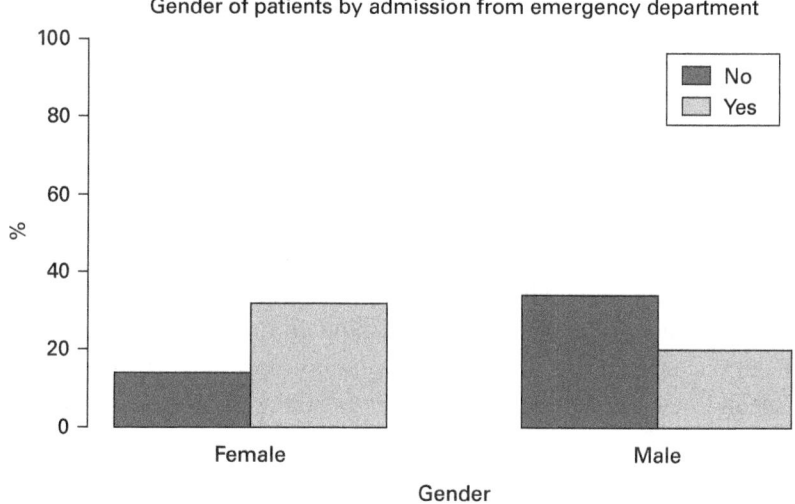

FIGURE 2.9. Example of side-by-side bar plot.

Looking at the bars side by side, it is now easy to see that a higher percentage of females are admitted to the hospital through the emergency department, whereas males are more likely NOT to be admitted through the emergency department.

## Mosaic Plots

Mosaic plots are used less commonly than bar plots, but they are aesthetically pleasing and can powerfully convey information about variables. As an example, we can use the *admit* table we created previously to visualize hospital admission by gender in another way.

To run a mosaic plot, we are going to use the *vcd* package function **mosaic()**. Be sure to download and load the *vcd* package prior to entering the **mosaic()** function. Then, enter the following into the Console to produce the graphic you see in Figure 2.10:

```
> mosaic(hospital$gender~hospital$edadmit)
```

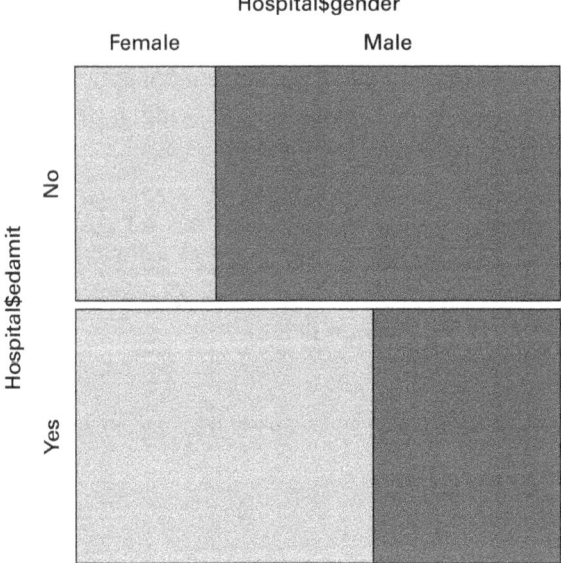

FIGURE 2.10. Mosaic plot of gender by hospital admission.

As you can observe from the mosaic plot in Figure 2.10, the squares are divided into horizontal bars with lengths proportional to their probabilities linked with the first categorical variable, admission through the emergency department. Each bar is split further so that they are proportional to the probabilities of the second categorical variable, gender of patients. The size of each tile is relative to the total proportion for the sample. Therefore, you can visually see that the smallest group is females not admitted through the emergency department, and the next smallest is males who are admitted through the emergency department.

## CONTINUOUS DATA

### Grouping Interval- and Ratio-Level Data and Establishing Percentiles

Unlike nominal or ordinal data, continuous data can have an infinite number of categories. For example, consider the ages of 10 people:

82, 56, 47, 89, 48, 88, 62, 27, 80, 32

We can think of describing this data in several ways. One way to do this would be to look at how many people were in their 20s, their 30s, their 40s, and so on. This data is displayed in Table 2.4.

TABLE 2.4. Grouping of People Into 10-Year Age Ranges

| Age Range | N | Cumulative Frequency | Cumulative Percent |
|---|---|---|---|
| 20–29 | 1 | 1 | 10% |
| 30–39 | 1 | 2 | 20% |
| 40–49 | 2 | 4 | 40% |
| 50–59 | 1 | 5 | 50% |
| 60–69 | 1 | 6 | 60% |
| 70–79 | 0 | 6 | 60% |
| 80–89 | 4 | 10 | 100% |

Notice that the age range categories that we created are both mutually exclusive and exhaustive; that is, age ranges do not overlap in any way and we created enough categories that we included all available data. Notice also that the cumulative frequency column includes the number of observations from the current age range *and* each previous age range. For example, the cumulative frequency for those in the 50- to 59-year age group includes all observations/individuals from the 20 to 29 age group, the 30 to 39 group, the 40 to 49 group, and the 50 to 59 group. The last value in the cumulative frequency column, the one for the 80 to 89 age group, includes all 10 individuals in this small dataset. Similarly, the cumulative percentage does the same, but on a percentage basis, and is the percentile rank for any observation. Therefore, we can quickly see that half the sample is aged 59 or younger, and those in the 50- to 59-year-old age group are in the 50th percentile. This represents the median, or middle value, which we will discuss later in this chapter.

Of course, larger datasets would be more difficult to sort by hand the way we did in Table 2.4, but luckily, we can do this easily in *R*. To do this, we will use the *fdth* package to improve our interpretation.

To illustrate this, we can look at the age of patients in the *hospital* dataset by entering the following in the Console:

```
> summary(hospital$age)
   Min. 1st Qu.  Median    Mean 3rd Qu.    Max.
  18.00   46.25   58.50   58.58   69.00  105.00
```

Notice that the youngest patient (*Min.*) is 18 and the oldest is 105 (*Max.*). *R* automatically calculates the median for you, which is 58.50. Now we can use the **fdt()** function to create categories as we did previously:

```
> f<-fdt(hospital$age,start=15,end=110,h=5,right=TRUE)
> f
```

The output is presented in Figure 2.11.

| Class limits | f | rf | rf(%) | cf | cf(%) |
|---|---|---|---|---|---|
| (15,20] | 1 | 0.02 | 2 | 1 | 2 |
| (20,25] | 0 | 0.00 | 0 | 1 | 2 |
| (25,30] | 0 | 0.00 | 0 | 1 | 2 |
| (30,35] | 0 | 0.00 | 0 | 1 | 2 |
| (35,40] | 6 | 0.12 | 12 | 7 | 14 |
| (40,45] | 5 | 0.10 | 10 | 12 | 24 |
| (45,50] | 4 | 0.08 | 8 | 16 | 32 |
| (50,55] | 6 | 0.12 | 12 | 22 | 44 |
| (55,60] | 7 | 0.14 | 14 | 29 | 58 |
| (60,65] | 5 | 0.10 | 10 | 34 | 68 |
| (65,70] | 4 | 0.08 | 8 | 38 | 76 |
| (70,75] | 2 | 0.04 | 4 | 40 | 80 |
| (75,80] | 6 | 0.12 | 12 | 46 | 92 |
| (80,85] | 3 | 0.06 | 6 | 49 | 98 |
| (85,90] | 0 | 0.00 | 0 | 49 | 98 |
| (90,95] | 0 | 0.00 | 0 | 49 | 98 |
| (95,100] | 0 | 0.00 | 0 | 49 | 98 |
| (100,105] | 1 | 0.02 | 2 | 50 | 100 |
| (105,110] | 0 | 0.00 | 0 | 50 | 100 |

FIGURE 2.11. Table of patients' ages by groups.

In this function, the **start** option indicates the lowest value for the first category, **end** indicates the highest value for the last category, and **h** indicates the size of each category. In this example, we know that the youngest person is 18 years of age, so a low value of 15 would include that individual. Similarly, a high value of 110 would include the oldest individual, who is 105. We indicated that we want ages to be incremented by five years, and we notice that the first interval includes people aged 15 to 20.

The column labeled *f* displays the frequencies for each column, and *rf* shows the proportions for these. *rf(%)* displays this information as a percentage, while *cf* shows the cumulative frequency, discussed earlier; *cf(%)* shows this information as a percentage. For example, 12% of the sample ($n = 6$) is between 50 and 55 years of age, and these individuals are in the 46nd percentile for age, meaning that 54% of the people in the sample are older than they are (100% − 46% = 54%).

### Displaying Continuous Data

You can visually depict continuous data in a variety of ways including histograms and stem-and-leaf plots, which will be discussed in this section.

The histogram can be employed when there is a need to display the distribution of a continuous variable. The following *R* code displays the ages of patients from the *hospital* dataset in Figures 2.12 and 2.13:

```
> hist(hospital$age, main="Histogram of Age",
xlab="Patient Age")
> hist(hospital$age,breaks="FD",col="lightgray",xlab=
"Patient Age",main="Histogram of Age" )
```

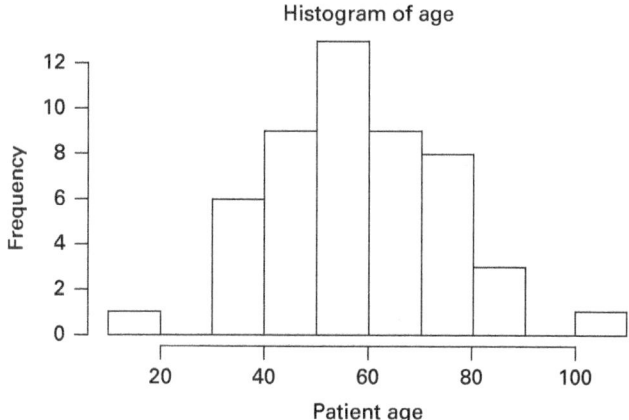

FIGURE 2.12. Histogram of patient age.

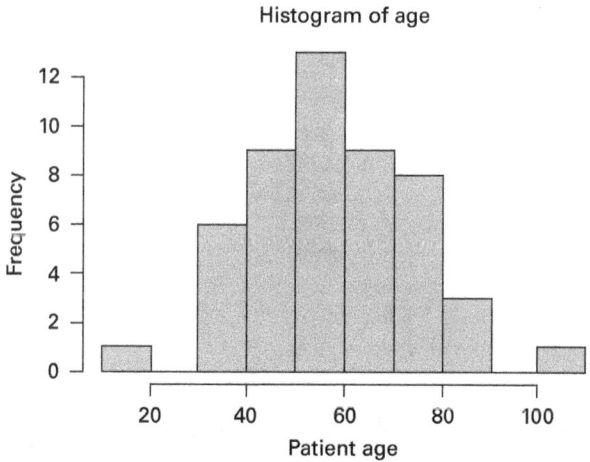

FIGURE 2.13. Enhanced histogram of patient age.

Notice how the histogram looks a lot like the bar graphs we produced earlier; however, in the case of the histogram, the bars, or bins as they are called in histograms, touch each other to denote the continuous nature of the data.

In the aforementioned functions, we created the first histogram simply. We allowed *R* to set the spacing of the bins, and we simply gave the histogram a main title and labeled the x-axis. In the enhanced histogram (Figure 2.13), we colored the bars by adding the option `col="lightgray"`. The `breaks = "FD"` option sets the number of bins. The number of bins will affect the shape of the histogram. Too few bins may prevent revealing important characteristics of the data, while too may bins may lead to an inaccurate interpretation of the data (Fox & Weisberg, 2011). Fox & Weisberg (2011) suggests using the rule set by Freedman and Diaconis (1981) for setting the number of bins (Freedman & Diaconis, 1981; Fox & Weisberg, 2011). The formula for doing this uses a weighted range (i.e., the difference between the minimum and maximum values divided by the interquartile range, which will be discussed later in this chapter). The `breaks="FD"` option uses this formula for determining the optimal number of bins; therefore, it is our preference to group continuous data in this way.

An interpretation of Figures 2.12 and 2.13 indicates that there are a number of **outliers** at the extremes of the sample. This helps us better visualize what we saw in the output from the `fdt()` function. This is important to know because it can impact the type of analysis we conduct later.

A stem-and-leaf plot of age is another way to visualize the distribution of a continuous variable. It is easily produced in *R* with the following function:

```
>stem(hospital$age)
```

The results in the Console are displayed in Figure 2.14.

```
 1 | 8
 2 |
 3 | 67
 4 | 0000122336779
 5 | 11234467899
 6 | 00112256999
 7 | 3568
 8 | 0000111
 9 |
10 | 5
```

FIGURE 2.14. Stem-and-leaf plot of workers' ages.

In a stem-and-leaf plot, the value of each observation is displayed individually. In this case, the "stems" represent the tens and hundreds place for each observation, while each leaf represents the ones, or units, place for each. For example, in Figure 2.14, we can see that for people in their 70s in our sample, we have the following individuals' ages:

73, 75, 76, and 78

Notice that two leaves for the 40-year-olds have 7s, so we know that there were two 47-year-olds.

The display is character based and not as aesthetically appealing as a histogram; however, the plot looks like a histogram if it were rotated counterclockwise by 90 degrees. The stem-and-leaf plot brings us to the same conclusion that age is not normally distributed, like a bell curve, in this sample. It should be noted that stem-and-leaf plots are not as popular ways to visualize data as histograms are.

## Central Tendency

Nominal data are easily summarized using percentages and proportions. However, there are several types of measures that can be used to summarize continuous data that don't include groupings. The most common methods are central tendency and measures of dispersion utilized together. Measures of central tendency define what is "typical" in a distribution, while dispersion describes the degree of variability of the data.

There are several measures of central tendency that, under different circumstances, may yield different results. As such, it is important to understand the strengths and weaknesses of each.

### The Arithmetic Mean

The arithmetic mean is what is frequently referred to as the average. The formula for the mean is displayed as Formula 2.1 and can be defined as the sum of all values in an array divided by the total number of elements. As the formula expresses, the scores are summed and weighted by the number of cases.

For a population, the formula for the mean is written as such, displayed in Formula 2.1:

$$\text{Formula 2.1:} \quad \mu = \frac{\Sigma X}{N},$$

where $\mu$ is the population mean, $\Sigma X$ is the sum of all the elements, and $N$ is the number of elements. Similarly, the formula for the mean of a sample is shown in Formula 2.2:

$$\text{Formula 2.2:} \quad \bar{x} = \frac{\Sigma x}{n},$$

where $\bar{x}$ is the sample mean, $\Sigma x$ is the sum of all the elements, and $n$ is the number of elements. Notice that the actual formulae for the population and sample means are the same except for the notation. As stated in Chapter 1, population parameters are notated with Greek letters, while sample statistics use English letters.

The following is an example of how to calculate the mean. In this instance, we asked 10 students how many hours per day they spent on social media. The answers they gave are as follows:

2, 3, 10, 2, 2, 8, 1, 5, 3, 2

We can calculate the mean length of time on social media as follows:

$$\bar{x} = \frac{\Sigma x}{n} = \frac{2+3+10+2+2+8+1+5+3+2}{10} = \frac{38}{10} = 3.8 \text{ hours}$$

Notice, again, how cumbersome it would be to manually calculate the mean with larger amounts of data. In this case, *R* can come in handy, and we could quickly calculate the mean age of patients from the *hospital* dataset with the following function entered into the Console:

```
> mean(hospital$age, na.rm=TRUE)
[1] 58.58
```

The inclusion of the option **na.rm=TRUE** tells *R* that we do not want to include missing values in our calculation of the mean, and the output tells us that the average age of patients is 58.58 years.

One of the drawbacks of the mean is the influence of outliers upon it. This becomes more apparent when the number of cases is small. Let's take, for example, the following hypothetical yearly salaries of a sample of people: $30,000, $28,200, $40,100, $29,945, $36,345, $41,200, and $734,324. The mean of these salaries is calculated as follows:

$$\bar{x} = \frac{\Sigma x}{n} = \frac{30,000 + 28,200 + 40,100 + 29,945 + 36,345 + 41,200 + 734,324}{7}$$
$$= \$134,302$$

In this distribution, the mean would not be a very good representation of the typical salary. Except for the salary of $734,324, all other values are well below $100,000, and we may want to look at another way of describing a typical salary for this sample.

### The Median

The median (Md) is a positional measure that is the value in a distribution that has the same number of observations below it as above it; that is, the median represents the middle value, or 50th percentile, when all the values of the observations are sorted in ascending order. If we take the salary data earlier and arrange it from lowest to highest salary, we can easily observe this value, which has the same number of scores above it as below it:

$28,200   $29,945   $30,000   $36,345   $40,100   $41,200   $734,324

The median value here is $36,345 since half the observations are below it and half are above it. That is, $36,345 represents the fourth of seven total salaries. Notice that the median salary is more typical than the mean of $134,302 since all observations are given equal weight. One of the advantages of the median is that, unlike the mean, it is not distorted by extreme values, either low or high. Being a positional measure based on a single value, however, it is not as robust a measure as the mean, which uses the actual value of all observations.

In the example of salary, there are an uneven number of observations, making it easy to discern the position of the salary that divides the scores evenly. This is not the case with values that have an even number of observations. When this occurs, the mean of the two most center scores is the median.

To illustrate this, let's look at the time spent on social media from our previous discussion on the arithmetic mean. Table 2.5 sorts these observations from low to high by number of hours with a position of 1 showing the fewest hours and a position of 10 showing the greatest number of hours.

TABLE 2.5. Hours Spent on Social Media for 10 Individuals

| Hours Spent on Social Media | Position |
|---|---|
| 1 | 1 |
| 2 | 2 |
| 2 | 3 |
| 2 | 4 |
| 2 | 5 |
| 3 | 6 |
| 3 | 7 |
| 5 | 8 |
| 8 | 9 |
| 10 | 10 |

In Table 2.5 the median is the mean of the fifth position at two hours and sixth position at three hours because, together, they are the center of the distribution. You can also locate the center point by using the following formula:

Formula 2.3:   $Md = 0.5(n+1) = 0.5(10+1) = 5.5$

This tells us the median is located between positions 5 and 6.

The median would be calculated as follows:

$$Md = \frac{(2+3)}{2} = 2.5$$

Again, figuring out the median by hand-calculating can be cumbersome, particularly with larger datasets, and we can use *R* to obtain the median easily. We can determine the median age of patients from the *hospital* dataset by entering the following into the Console:

```
> median(hospital$age, na.rm=TRUE)
[1] 58.5
```

Notice that the median age of 58.5 years is close to the mean age of 58.58 years. This is because there are relatively few outliers compared to the sample size.

### Trimmed Mean

The trimmed mean (tM) can also be used when there are outliers because this measure of central tendency can reduce their impact (Auerbach & Zeitlin, 2014; Bloom, Fischer, & Orme, 2009; Verzani, 2004). The trimmed mean removes a percentage of the highest and lowest values when calculating the mean, thus "trimming" the total number of observations used. A common practice is to exclude 10% of the lowest and highest values to eliminate the influence of outliers at both the high and low ends. The tM is often preferable to the median since more information about the data is included in the calculation. However, if the number of observations is too small, the trimmed mean should not be utilized. Furthermore, if there are no extreme values (i.e., outliers), the mean would be preferable.

Here are the steps to calculate the 10% trimmed mean for the social media usage data in Table 2.5:

| Step | Calculation |
|---|---|
| Calculate how many scores are 10% of the total. This will be the number of observations to remove from both the high and low values. | $0.10 \times n$<br>$0.10 \times 10 = 1$ |
| Remove one observation from the lowest values and one observation from the highest values | 10% of the lowest values (1) = 1<br>10% of the highest values (1) = 10 |
| Calculate the sum ONLY using the remaining values | $10\% \text{ tM} = \dfrac{2+2+2+2+3+3+5+8}{8}$<br>$= 3.375 \text{ hours}$ |

To calculate the 10% trimmed mean for the age of patients in the *hospital* dataset, enter the following syntax into the Console:

```
> mean(hospital$age,trim=.1,na.rm=T)
[1] 58.25
```

Notice that the addition of the option of `trim=0.1` to the `mean()` function removes 10% of the values from the lowest and highest scores. The value for the `trim` option can range from 0 to 0.5, which would be a 50% trimmed mean, leaving no data to calculate a mean from since you would be removing the highest 50% of observations and the lowest 50% of observations.

Since there are few outliers in the age of patients, our true mean of 58.58 years is very close to the 10% trimmed mean of 58.25 years, and use of the actual mean is preferred.

### The Mode

The mode is simply the most frequently occurring value or score across all observations. For example, in Table 2.5, the modal hours of social media use is two hours, with four individuals falling into that category. Unlike the other measures of central tendency discussed previously, it is possible to have multiple modes (e.g., bimodal); therefore, the mode is not a popular measure of central tendency as it is difficult to discern a "typical" value.

## Measures of Dispersion

While measures of central tendency describe what is typical for a given set of data, these calculations tell us nothing about how data are dispersed. For example, consider the following dataset, which represents the number of children in five different families:

1, 2, 3, 4, 5

We can easily calculate the mean and median for this group of families, which is 3. Now consider this dataset, which represents the number of children in another five families:

3, 3, 3, 3, 3

Again, both the mean and median for this group is 3; however, the data in each of these datasets is completely different despite 3 being a good representation of the typical value for each.

Because of this, measures of dispersion are commonly reported along with measures of central tendency to further describe the spread of data about some reported typical value. Common measures of dispersion include range, interquartile range, standard deviation, and variance.

### The Range

The range is the simplest measure of dispersion, and it is the difference between the highest (maximum) score and lowest (minimum) score. One advantage of being a very simple measure is that people can easily comprehend its meaning. It also provides a very quick view of dispersion. Given the range's simplicity and use of only two values, we recommend reporting both the minimum and maximum values when using the range; however, we suggest generally using more robust measures of dispersion.

If we think about the two samples of families, described earlier, we could now get a better picture of the data by noting that the first sample had a mean of 3 and a range of 4, while the second sample also had a mean of 3, but that sample had a range of 0, indicating no variability among the number of children in each family.

In the example of the time spent on social media, the range of that data would be $10 - 1 = 9$ hours. With large datasets, calculating the range could be somewhat time consuming because you would need to look through all the data to identify the minimum and maximum values. Using the **describe()** function from the *psych* package in *R* is a simple way to obtain the range along with other summary statistics for a single variable. To obtain the results illustrated in Figure 2.15, enter the following into the Console after loading the *psych* package:

```
> describe(hospital$age)
```

```
      vars  n  mean    sd median trimmed   mad min max range skew kurtosis   se
  X1     1 50 58.58 16.43   58.5   58.25 17.79  18 105    87 0.24    -0.05 2.32
```
FIGURE 2.15. Summary information for patients' ages.

The output from Figure 2.15 displays a lot of information that we learned about already. First, we see that there are 50 observations for this variable, and the mean, median, and trimmed mean are also shown, with the trimmed mean being labeled *trimmed*. We can see the minimum and maximum values along with the range, which is 87.

### The Interquartile Range

The interquartile range (IQR) is the difference between the third quartile, or the 75th percentile, and the first quartile, or 25th percentile, and it denotes the middle 50% of all observations for a given variable. To calculate this, we can use the data we described in Table 2.5 to determine both the 25th and 75th percentiles. In Table

2.6, however, note that how often a given value appears in the data is notated in the relative frequency column.

TABLE 2.6. Calculating Interquartile Range for Social Media Usage

| Hours Spent on Social Media | Relative Frequency (rf) | Cumulative Frequency (cf) | Cumulative Percentage (cf%) |
|---|---|---|---|
| 1 | 1 | 1 | 10 |
| 2 | 4 | 5 | 50 |
| 3 | 2 | 7 | 70 |
| 5 | 1 | 8 | 80 |
| 8 | 1 | 9 | 90 |
| 10 | 1 | 10 | 100 |

Notice how Table 2.6 looks similar to the one displayed in Figure 2.11, which showed output from the *R* `fdt()` function.

We can see that a value of 2 represents the 25th percentile since a value of 1 only reaches the 10th percentile. Similarly, we can see that a value of 2 also represents the median, which is the 50th percentile. The 75th percentile is much more difficult to determine since it falls somewhere between a value of 3 and a value of 5, but there is no value of 4 in the dataset. Determining the exact value for the 75th percentile here requires linear interpolation, which is difficult to calculate by hand; however, *R*'s `quantile()` and `IQR()` functions make this simple.

To illustrate this, we will begin by building a vector to hold the social media information and then we will use these functions to display both the quartiles and the interquartile range. Enter the following into the Console:

```
> sm<-c(1, 2, 2, 2, 2, 3, 3, 5, 8, 10)
> quantile(sm)

  0%  25%  50%  75% 100%
 1.0  2.0  2.5  4.5 10.0

> IQR(sm)
[1] 2.5
```

Notice that the IQR has the same value as the median, which is indicative of a skewed distribution. A box plot of social media use is helpful in visualizing this. The following code was used to display the box plot in Figure 2.16:

```
>boxplot(sm, main="Social Media Use in Hours",ylab="H
ours",col="lightgray")
```

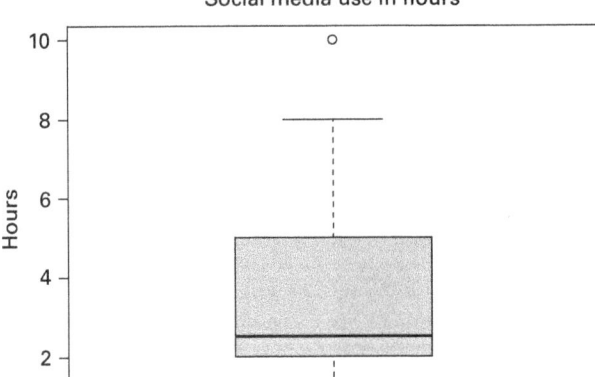

FIGURE 2.16. Box plot of social media usage.

The dark black line inside the box represents the median, and circles are outliers, defined in *R* as data points beyond 1.5 times the interquartile range. The bottom of the box itself represents the 25th percentile, while the top of the box represents the 75th percentile. The box plot shows that the data are skewed to the lower values.

Because the IQR includes information about the variability among extreme values, it is a more stable measure of variability than the range, but it still does not take advantage of all the observations since it excludes the bottom and top 25% of data.

### The Standard Deviation and Variance

The most frequently used measure of dispersion is the standard deviation. It can be defined as the average amount of variability of the scores from the arithmetic mean. The formula for this is displayed in Formula 2.4, which represents the standard deviation of a population, as denoted by the Greek symbol σ:

$$\text{Formula 2.4:} \quad \sigma = \sqrt{\frac{\Sigma(X-\mu)^2}{N}}$$

The sample standard deviation is calculated slightly differently:

$$\text{Formula 2.5:} \quad s = \sqrt{\frac{\Sigma(x-\bar{x})^2}{n-1}}$$

Notice that the only real difference in these formulae is in the denominator of the equation. *N* is used in the formula for the population, while *n* − 1 is used in the formula for the sample. The reason for this is that when samples are small, a value of *N* in the denominator of the equation for *s* would underestimate the variability of the data as a predictor for the true population standard deviation, or σ. Therefore, $n - 1$ serves a correction for the unbiased standard deviation.

For more information on why we use n–1 in the sample calculations for standard deviation and variance, watch the KHAN ACADEMY® videos "Review and intuition why we divide by n–1 for the unbiased sample variance "

To calculate the standard deviation for social media usage, begin by subtracting the mean (3.8 hours) from each observation. The next step is to square the difference and add those values, which is called the sum of squares (SS). Next, divide by $n - 1$ (since this data is based on a sample) and, finally, take the square root. These calculations are shown in Table 2.7.

TABLE 2.7. Calculating $s$ for Patients' Length of Stay

| Individual | Social Media Use (hours) | $x - \bar{x}$ | $(x - \bar{x})^2$ |
|---|---|---|---|
| 1 | 2 | −1.8 | 3.24 |
| 2 | 3 | −0.8 | 0.64 |
| 3 | 10 | 6.2 | 38.44 |
| 4 | 2 | −1.8 | 3.24 |
| 5 | 2 | −1.8 | 3.24 |
| 6 | 8 | 4.2 | 17.64 |
| 7 | 1 | −2.8 | 7.84 |
| 8 | 5 | 1.2 | 1.44 |
| 9 | 3 | −0.8 | 0.64 |
| 10 | 2 | −1.8 | 3.24 |
| $\bar{x} = 3.8$ | $\Sigma x = 38$ | $\Sigma x - \bar{x} = 0$ | $\Sigma (x - \bar{x})^2 = 79.6$ |

$$s = \sqrt{\frac{79.6}{9}} = \sqrt{8.844} = 2.9738$$

Notice that the sum of the difference of each observation from the mean is equal to zero, which is always the case. Squaring these values, however, eliminates negative numbers and enables an accurate calculation of the standard deviation. In this case the sum of squares is equal to 79.6, and the standard deviation is 2.9738.

The variance has a close relationship to the standard deviation. The definition is shown in Formulae 2.6 and 2.7 for both a population and a sample:

$$\text{Formula 2.6}: \quad population\ variance = \sigma^2 = \frac{SS}{N} = \frac{\Sigma(X - \mu)^2}{N}$$

$$\text{Formula 2.7:} \quad \textit{sample variance} = s^2 = \frac{SS}{n-1} = \frac{\Sigma(x-\bar{x})^2}{n-1}$$

The only difference in computation of the variance and the standard deviation is the removal of the square root. In order to do this, we simply square both sides of the equation. Therefore, the symbol for the population variance is $\sigma^2$ and the symbol for the unbiased sample variance is $s^2$. In Table 2.7, we can easily calculate the variance for patient's length of stay, which is 8.844.

In the previous example, if every individual's social media usage had been the same (say, one hour), the standard deviation would have been zero. When interpreting standard deviation or variance, you will notice that the larger the value in relation to the mean, the greater is the spread of the data around the mean. Also notice that the scores with the greatest deviations, either positive or negative, have the most influence on the standard deviation or variance. In Table 2.7, then, observations from individuals 3 and 6 are the most influential. Like the mean, if the number of observations is small and there are a few large deviations, the standard deviation can be misleading.

In general, the standard deviation is a much more useful and interpretable measure of variability than the variance, and it is often reported in conjunction with the mean to give a comprehensive depiction of a set of data. The usefulness of the variance will become more apparent later in the book when analysis of variance is discussed.

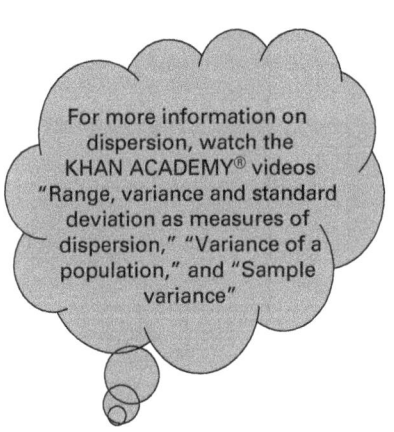

For more information on dispersion, watch the KHAN ACADEMY® videos "Range, variance and standard deviation as measures of dispersion," "Variance of a population," and "Sample variance"

You may have noticed how tedious it is to calculate the standard deviation and variance with only 10 observations; larger amounts of data would be even more cumbersome and would introduce the potential for even more error in manual calculations. We therefore recommend using *R* to increase accuracy and speed.

To calculate the standard deviation of the ages of all 50 patients in the *hospital* dataset, enter the following into the Console:

```
> sd(hospital$age, na.rm=T)
[1] 16.42806
```

To calculate the variance, you would use the **var ()** function:

```
> var(hospital$age, na.rm=T)
[1] 269.8812
```

Alternately, output from the **describe()** function, illustrated in Figure 2.15, displays the standard deviation under the *sd* heading.

## The Coefficient of Variation

Earlier, we mentioned that the standard deviation is interpreted relative to the mean. Consider a standard deviation of $1,000. On the face of it, this may seem like a very high value, but we need to look at it in context. If we were talking about the price of computers, this standard deviation would seem high if the mean cost of a computer were $1,200. If, on the other hand, we were talking about annual salaries with a mean of $90,000, a standard deviation of $1,000 would indicate very little variability.

One way to quantify this and compare these is with the coefficient of variation (CV), which is defined in Formula 2.8 as:

$$\text{Formula 2.8}: \quad CV = \frac{standard\ deviation}{mean}$$

With this information, we can calculate the CV for computer prices:

$$CV = \frac{standard\ deviation}{mean} = \frac{1,000}{1,200} = 0.8333$$

We can also calculate the CV for the salary information:

$$CV = \frac{standard\ deviation}{mean} = \frac{1,000}{90,000} = 0.0111$$

Now, we can clearly understand that the salary data is much less variable about the mean than computer prices since the CV for computer prices is much higher than it is for salary.

## CONCLUSION

In this chapter, we looked at ways to describe both categorical and continuous data. Categorical data are best described using proportions and percentages. We can visualize these data easily using bar plots. When describing continuous data, measures of central tendency and dispersion are preferable. Measures of central tendency include mean, median, and mode, and measures of dispersion include the range, interquartile range, standard deviation, and variance. Again, visually depicting continuous data helps in the

For more information on when to summarize data with mean and standard deviation compared to median and IQR, watch the KHAN ACADEMY® video "Mean and standard deviation versus median and IQR"

interpretation. Histograms, stem-and-leaf plots, and box plots are all methods for doing this effectively.

## EXERCISES

For each of the following, please use the *student* data, located in Appendix D.

1.
   a. Calculate the mean, median, and standard deviation for students' GPA scores.
   b. Based on your calculation, does the mean adequately represent this sample? Why or why not?
   c. How would the mean and median of GPA be impacted if the lowest score is removed?
   d. What would be the effect on the distribution of GPA if the two lowest scores and the two highest scores were removed? Describe this new sample of data distribution in terms of the mean, median, and standard deviation.
2. Calculate the interquartile range for GPA.
3.
   a. Calculate the coefficient of variation for the GPA scores.
   b. What does this tell you about variation in this variable?
4. Provide the frequencies and percentages of females and males for the students in the sample.

## *R* EXERCISES

For the following questions, use the *hospital* dataset included with this text and described in Appendix D.

1. What is the mean, median, and standard deviation of the *sensitivity* variable?
2. What is the 10% trimmed mean of the *sensitivity* variable?
3. Create a box plot of the *sensitivity* variable. What does the box plot tell you about the distribution? Based on this information, do you think it is better to use the mean, trimmed mean, or median as a description of what is typical for this variable?
4. Create a cumulative frequency table for the variable *length of stay (LOS)*, grouping the data by intervals of 5. What is the largest category for LOS?
5. Create a table to show what the percentages of males and females are in this sample. Now create and label a bar plot to show this same information.

# PROBABILITIES AND THE BINOMIAL DISTRIBUTION

## INTRODUCTION

In this chapter, we are going to begin by thinking about the chances of an event occurring. When we think about systematic events, such as those we will be discussing here, we can begin thinking about cause-and-effect relationships. If there is a cause-and-effect relationship, the effect will be dependent upon the cause in some way. Additionally, there can be other things that we observe that may suggest a relationship between two variables. If we do not believe that two events are related in some way, then we say that those events are independent of one another.

Furthermore, we can consider the different ways that data can be distributed based on the probability of an event happening. When we contemplate this, we can think about what constitutes usual events compared to unusual events.

Later in this chapter, we will introduce the binomial distribution, which is the first of several distributions we will talk about in this book. You will come to learn that distributions can be used to help better understand when an observation can be considered usual compared to being unusual.

## PROBABILITY: FLIPPING COINS, ROLLING DICE, AND DEALING CARDS

In thinking about probability, we want to consider the chances of an event happening that could occur randomly. Subjective probability uses previous experience to think about whether an event may occur. For example, if you look up at the sky and see clouds, you could conclude that it is likely to rain. This conclusion is based on your previous experience of it raining under similar conditions; however, it may not rain, and you would not be able to predict with any degree of certainty whether it is actually going to rain without a lot more information.

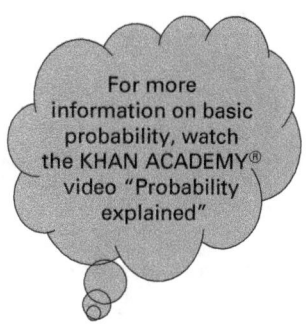

For more information on basic probability, watch the KHAN ACADEMY® video "Probability explained"

Objective probability is a bit more predictable and will be what we consider here. An example of this is a coin that was flipped once has a 50% chance of landing on heads. This idea of probability can be extended to predicting relative frequencies, which we discussed in the last chapter. If we extend the coin-flipping example and move on to flipping the coin 1,000 times, based on the probability of getting a head 50% of the time, we could predict that we would get heads 500 times:

$$1{,}000 \text{ coin flips} \times 0.50 = 500 \text{ heads}$$

If we were going to do the experiment of flipping a coin 1,000 times, we may not get heads 500 times, but the result is likely to be pretty close to this.

Looking at this seemingly intuitive conclusion, we can begin discussing some important basic concepts related to probability. We can consider what the possible events are surrounding a single coin flip. An event is notated in statistics with *E*. One outcome of the coin-flipping event is that the coin could land on heads, and the other outcome is that it could land on tails. The sample space contains all possible outcomes:

$$S_C = \{H, T\},$$

where $S_C$ represents the sample space (S) for coin flips (C). Similarly, we could define the sample space for the roll of a single die as:

$$S_D = \{1, 2, 3, 4, 5, 6\}$$

since the roll of a single die could yield a 1, 2, 3, 4, 5, or 6 and nothing else.

Furthermore, *p* denotes the probability of an event happening. The probability of any particular event occurring is shown in Formula 3.1:

$$\text{Formula 3.1:} \quad p(E) = \frac{\# \text{ of successes}}{\text{Total number of possible outcomes}}$$

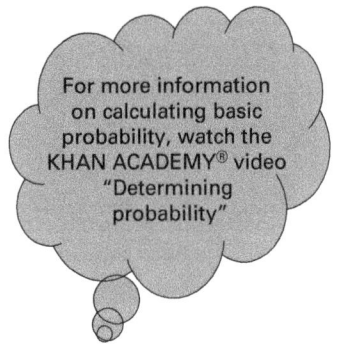

For more information on calculating basic probability, watch the KHAN ACADEMY® video "Determining probability"

In the coin-flipping example, we defined "success" as obtaining a head out of two possible outcomes (heads or tails). Therefore, the probability for obtaining a head is:

$$p(head) = \frac{\# \text{ of successes}}{\text{Total number of possible outcomes}}$$
$$= \frac{1}{2} = 0.50$$

(While you may have known this intuitively, this is the math behind that intuition. It will come in handy later.)

The nature of probability in general is based on some basic rules:

1. $p(E) > 0$

The probability of any event/outcome in the sample space occurring in an experiment is greater than zero. When we think of rolling a die, for example, it would not be impossible to, say, get a 6. There is always some chance of obtaining any of the outcomes in a given sample space.

2. $0 < p(E) \leq 1$

This extends the previous rule somewhat. While we know that the probability of any single event occurring is greater than zero, we also know that the probability of that same event occurring has to be less than or equal to one, which would indicate a 100% chance of that event occurring.

3. $\sum p(E) = 1$

This means that the sum of the probabilities of all possible events happening in an experiment must be 1, representing all, or 100%, of the possible outcomes. In the coin-flip example, we know that the probability of getting a head is 0.50, or 50%. Similarly, the probability of getting a tail is also 0.50. There are no other possible outcomes, and 0.50 + 0.50 = 1.

For good examples of calculating basic probability, watch the KHAN ACADEMY® videos "Finding probability example," "Finding probability example 2," "Finding probability example 3"

## A Little More About Success

In the coin-flip example, we defined success as getting a head. Notice that success is not necessarily defined as something positive, but as the outcome in which you were most interested. Success in statistical terms, then, can be something either intuitively positive, negative, or even neutral; it is simply the outcome of interest.

The statistical notation for the number of successful outcomes in an experiment is $k$. The total number of outcomes in an experiment is denoted as $n$. Therefore, the total number of failures is defined as $n - k$, as failure is defined as any event other than what is considered a success. While the probability of a successful event occurring is notated as $p$, the probability of failure is defined as:

$$1 - p = q$$

Now that we understand this notation, we can add to our basic rules that we began previously:

4. $0 < p(n-k) \leq 1$

Similar to the second rule, this one tells us that the probability of failure in an experiment is always greater than zero and less than or equal to one. There is never NO chance of complete failure, but there is also never the possibility that the probability of failure is greater than one, or 100%.

### Some More to Consider

Before we move on to examples that are more complicated than our single coin flip, we need to think about two concepts that also influence probability when we are interested in more complex definitions of success in an experiment. They are **mutual exclusivity** and **independence**.

When we think about events being mutually exclusive, we are considering whether there is potential overlap between possible outcomes. For example, let us consider drawing a card at random from a standard deck of 52 cards. If we were interested in whether the card we drew was a club or a diamond, we know that there

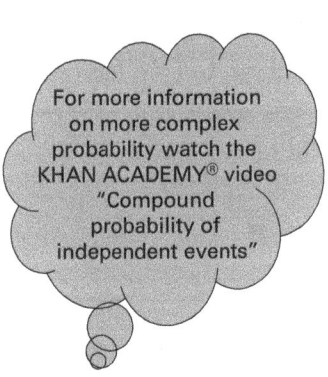

For more information on more complex probability watch the KHAN ACADEMY® video "Compound probability of independent events"

is no possible overlap in these possible outcomes because, by definition, no clubs in a deck are also diamonds and no diamonds in a deck are also clubs. However, if we defined success as being either a club or an ace, we would have to consider possible overlap because there is a single card, the ace of clubs, which would fit the criteria for success in both the suit category and the number category. Therefore, when we calculate probabilities with more complicated definitions of success, we will need to treat mutually exclusive outcomes differently than those that are not mutually exclusive.

Independence is another factor to consider in calculating probabilities and needs to be thought about when there is more than one event occurring in an experiment. We need to ask, "Are each of the events in the experiment truly independent of one another or are they related in some way?" As an example, let us think about rolling two dice (or, alternatively, rolling one die twice). Is the outcome of one somehow dependent upon the other? No. Whatever happens on the roll of one die has no influence whatsoever on what happens when the second die is rolled. The same is true when you flip coins.

Drawing more than one card from a deck, however, is a different matter. If we are interested in drawing two cards from a deck, the question becomes whether you replace the first card before drawing the second card. If you draw one card and then replace it, it is as if you have two independent experiments, as the second draw will have all 52 cards in the deck in the same way as the first draw. If, however, you do not replace it, one card will be missing from the deck (i.e., the one you drew first), and you will only have 51 cards remaining in the deck. Therefore, the results of the second draw in this instance will be related to what happened in the first draw.

When there is a single event in an experiment and there is no chance of overlap (i.e., the possible outcomes are mutually exclusive), we can think of probability like this:

$$\text{Formula 3.2}: \quad p(A \text{ or } B) = p(A) + p(B)$$

To illustrate this, consider the most basic example again of a simple coin flip. We could calculate the probability of getting a head or a tail as follows:

$$p(\text{head}) = 0.5 \text{ and } p(\text{tail}) = 0.5, \text{ and the } p(\text{head or tail}) = 0.5 + 0.5 = 1$$

This intuitively makes sense because there are only two possibilities in flipping a coin so a single flip will have to yield either a head or a tail.

Now consider rolling a single die and defining success as getting either a 1 or a 2:

$$p(1 \text{ or } 2) = p(1) + p(2)$$

We know that the probability of an event occurring was described in Formula 3.1:

$$p(1) = \frac{\# \text{ of successes}}{Total \text{ number of possible outcomes}} = \frac{1}{6} = 0.167$$

and

$$p(2) = \frac{\# \text{ of successes}}{Total \text{ number of possible outcomes}}$$
$$= \frac{1}{6} = 0.167$$

Watch Sal Khan talk more about rolling dice on the KHAN ACADEMY® videos "Die rolling probability" and "Die rolling with independent events"

Therefore, $p(1 \text{ or } 2) = 0.167 + 0.167 = 0.333$, or 33%.

Now we can consider a more complicated situation when we have more than one event. For example, in two coin flips, what would the probability be of getting two heads? As we said earlier, the results of flipping two coins (or even one coin twice) are independent events; what happens on the first flip has no impact on the result of the second flip assuming that your coin is a normal, fair coin. With independence in mind:

$$\text{Formula } 3.3: \quad p(\text{A and then B}) = p(\text{A}) \times p(\text{B})$$

Therefore, if we define "success" as getting a head on each roll, the probability of getting two heads consecutively is:

$p(\text{2 heads}) = p(\text{head}) \times p(\text{head}) = (0.5)\,(0.5) = 0.25 \text{ or } 25\%$

Watch Sal Khan talk more about flipping coins onthe KHAN ACADEMY® video "Coin flipping probability"

Similarly, we could calculate the probability of drawing an ace from a deck of cards on the first draw and then, replacing the first card, getting a diamond on the second draw.

To deal with this, we first need to know the probability of getting an ace and also the probability of getting a diamond.

Using Formula 3.1:

$$p(ace) = \frac{\#\ of\ successes}{Total\ number\ of\ possible\ outcomes}$$
$$= \frac{4}{52} = 0.077$$

and

$$p(diamond) = \frac{\#\ of\ successes}{Total\ number\ of\ possible\ outcomes} = \frac{13}{52} = 0.25$$

Therefore, assuming replacement of the first card before the second draw is made, which ensures that each draw is independent:

$p(\text{ace and then a diamond}) = p(\text{ace}) \times p(\text{diamond}) = (0.077)(0.25) = 0.0193$

The probability, then, of drawing two cards from a deck with the first being an ace and the second being a diamond is just under 2%.

Since the commutative law of multiplication lets us know that the order in which numbers are multiplied does not impact the results, we also know that, with replacement, it does not matter whether we are hoping to draw an ace first and then a diamond or a diamond first and then an ace since:

$$(0.077)(0.25) = (0.25)(0.077) = 0.0193$$

Now suppose we again want to draw two cards, but this time we will not replace the first card before drawing the second card. This time, we would like to know the probability of getting an ace and then getting a second ace. In this scenario, we add the complexity of nonindependence. If we are not replacing the first card that we drew, we need to consider the fact that the first event was successful. In this case, then:

$$\text{Formula 3.4:} \quad p(A \text{ and then } B) = p(A) \times p(B \mid A)$$

Notice the slight change in notation here from Formula 3.3. Here, $p(B \mid A)$ means "the probability of B given that A has occurred." We know that the probability of getting the first ace is 0.077 from the previous problem, but what is the probability of getting the second ace given that you got an ace on the first draw? Since we did not replace the first card we drew from the deck, we are left with 51 cards. Also, assuming that the first card was an ace, we also know that there are only three aces left in the deck. Therefore:

$$p(ace \mid ace) = \frac{\# \text{ of } successes}{Total \text{ } number \text{ } of \text{ } possible \text{ } outcomes} = \frac{3}{51} = 0.059$$

In this problem, then, the probability of drawing two aces in a row without replacement is:

$$p(ace \text{ and then } ace) = p(A) \times p(B \mid A) = (0.077)(0.059) = 0.005$$

Your chance of getting two aces in a row is very small—only half of 1%!

In these examples, all events were mutually exclusive. There was no overlap between possible outcomes, but consider this problem: In one draw, what is the probability of getting either an ace or a heart from a deck of cards? We know that there are four aces in a deck and 13 hearts, but there is also one card that overlaps both categories—the ace of hearts. In the case of nonmutually exclusive outcomes, we need to remove the possible overlap to avoid "double counting," in this case, the ace of hearts. Therefore, the following formula applies:

$$\text{Formula 3.5:} \quad p(A \text{ or } B) = p(A) + p(B) - p(A \text{ and } B)$$

Applying this to our problem, we have:

$$p(ace \text{ or } heart) = p(ace) + p(heart) - p(ace \text{ and } heart) = \frac{4}{52} + \frac{13}{52} - \frac{1}{52} = \frac{16}{52} = 0.308$$

In reality, Formula 3.5 applies to the situation in which there is mutual exclusivity; however, the $p$(A and B), the last part of the formula, is always zero, since that is actually the definition of mutual exclusivity. Therefore, it may be easier to consider these as two slightly different formulae.

## What's the Difference Between Probability and Odds?

Very often, people will interchange the idea of probability with odds, as in "What are the odds of getting an ace if I draw one card from a deck?" While probability and odds are related, they are slightly different, and their interpretations are different. Therefore, it is important at this point to differentiate the two because both are used in the study of statistics.

So far, we have looked at the probability of drawing an ace from a deck of cards:

$$p(ace) = \frac{\#\ of\ successes}{Total\ number\ of\ possible\ outcomes} = \frac{4}{52} = 0.077$$

Odds are defined differently:

$$\text{Formula 3.6:} \quad odds(ace) = \frac{\#\ of\ possible\ successes}{\#\ of\ possible\ failures} = \frac{4}{48} = 0.083$$

Notice that the difference between the formulae for calculating probabilities and odds lies in the denominator. Realize that the total number of possible outcomes is the sum of possible successes and possible failures. Therefore, the number of possible failures is equal to the difference between total possible outcomes and the number of possible successes. In the previous example, then, we determine the number of possible failures as 48 (52 cards in a deck less four possible successes because there are four aces in a deck).

Looking at a simpler example, consider the odds of getting a head on a single coin flip:

$$odds\ (head) = \frac{\#\ of\ possible\ successes}{\#\ of\ possible\ failures} = \frac{1}{1} = 1$$

In the coin-flip example, there are only two possible outcomes, heads or tails, leaving a one in both the numerator and denominator of the fraction. Therefore, an odds calculation of one is often referred to as "even odds."

## VISUALIZING PROBABILITIES

Consider flipping a coin twice and the fact that we might be interested in heads as a success. In flipping the coin twice, there are four possible combinations that could be made:

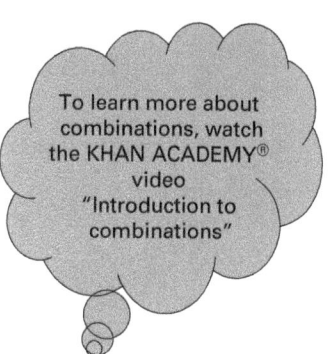

HH, HT, TH, TT

Notice that "HT" is not the same as "TH" because in the "HT" scenario we get a head on the first flip and a tail on the second flip; in the "TH" situation we get a tail on the first flip and a head on the second flip.

Also notice that, in terms of success, we have three scenarios. We get could no heads (TT), one head (HT or TH), or two heads (HH). The probability of all of these together, of course, is equal to one:

$$p(0\ H) = \frac{1}{4} = 0.25$$

$$p(1\ H) = \frac{2}{4} = 0.50$$

$$p(2\ H) = \frac{1}{4} = 0.25$$

Figure 3.1 displays a histogram of this.

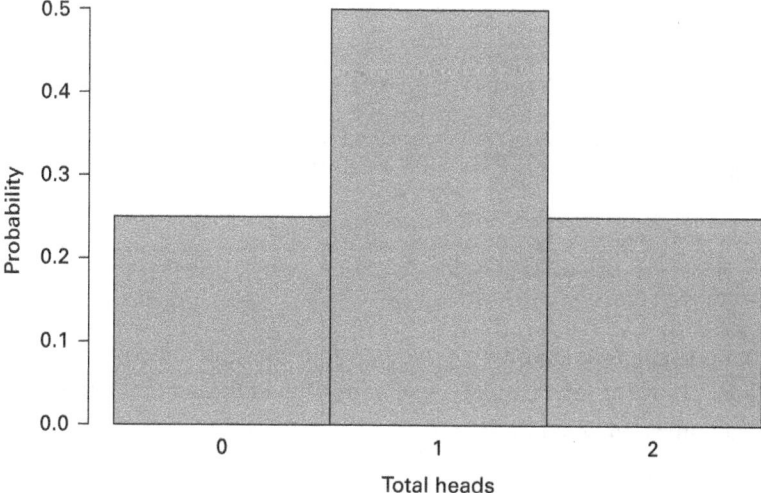

FIGURE 3.1. Plot of the probability of getting heads in flipping a coin twice.

If we flipped a coin three times, there are eight possible resulting combinations:

HHH, HTH, HTT, HHT, TTH, TTT, THT, THH

In terms of getting heads, there are now four scenarios: zero heads, one head, two heads, or three heads. Looking at possible outcomes, we can determine the probability of obtaining each of these:

$$p(0\ H) = \frac{1}{8} = 0.125$$

$$p(1\ H) = \frac{3}{8} = 0.375$$

$$p(2\ H) = \frac{3}{8} = 0.375$$

$$p(3\ H) = \frac{1}{8} = 0.125$$

Figure 3.2 displays a histogram of this.

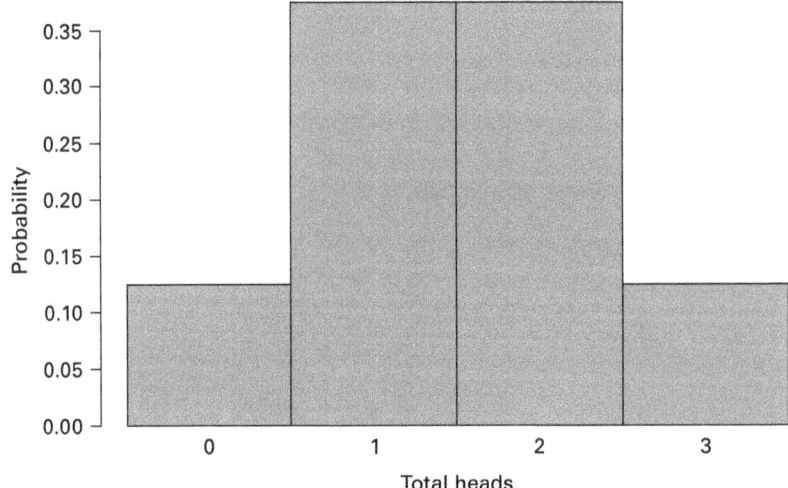

FIGURE 3.2. Plot of the probability of getting heads in flipping a coin three times.

When we flipped the coin twice, we had four possible outcomes, and when we flipped three coins (or, alternatively, one coin flipped three times), there were eight possible outcomes. In general:

Formula 3.7:   *Total possible outcomes*
                *= # possible out comes in each event$^{\#\,of\ events}$*

For each flip of the coin, there are two possible outcomes. When we flipped the coin twice:

$$Total\ possible\ outcomes = 2^2 = 4$$

and when we flipped the coin three times:

$$Total\ possible\ outcomes = 2^3 = 8$$

It may be apparent by now that identifying all possible outcomes in a more complicated experiment could be cumbersome. What would happen if we were interested in rolling five dice? We would have to identify $2^5$, or 32, outcomes to figure out the probability of success for each of six outcomes (zero heads, one head, two heads, . . . five heads). There must be an easier way and there is: the binomial distribution.

## THE BINOMIAL DISTRIBUTION

As we were considering our coin-flipping examples, we were interested in all the possible combinations that could result from our flipping experiment. Then, we categorized them by how many heads we got for each of the possible outcomes, since heads was our measure of success. Therefore, the first thing to consider in the binomial distribution is how many combinations satisfy each of the possible outcomes for success. With only three coin flips, that was not terribly difficult, but what would happen if we had more—a lot more?

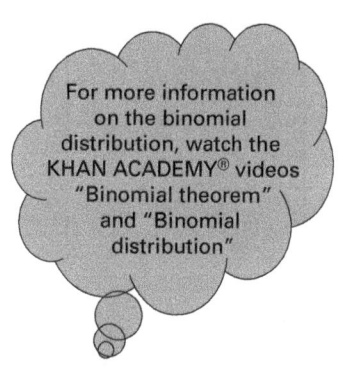

For more information on the binomial distribution, watch the KHAN ACADEMY® videos "Binomial theorem" and "Binomial distribution"

Luckily, there is an easier way to do this than simply listing everything manually and counting them up. In general:

$$Formula\ 3.8: \quad [_nC_k] = \frac{n!}{(k!)(n-k)!}$$

where $[_nC_k]$ denotes how many combinations of $n$ things you can make with $k$ successes.

As an example, let's think about flipping a coin four times with five possible outcomes—zero heads, one head, two heads, three heads, and four heads. If we wanted to know how many combinations could be made with an outcome of zero heads, we have:

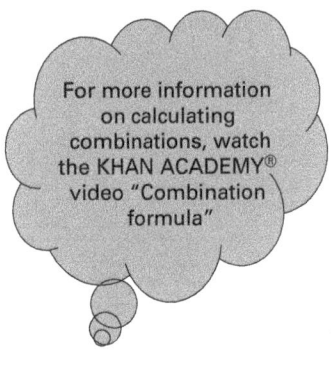

For more information on calculating combinations, watch the KHAN ACADEMY® video "Combination formula"

$$[_4C_0] = \frac{4!}{(0!)(4-0)!} = \frac{4!}{(1)(4)!} = 1 \text{ (which would}$$

be TTTT)

What about if we wanted to know how many combinations with one head could be made by make by flipping a coin four times?

$$[_4C_1] = \frac{4!}{(1!)(4-1)!} = \frac{(4)(3)(2)(1)}{(1)(3)(2)(1)} = 4$$

There are four combinations that can be made with only one head (HTTT, THTT, TTHT, and TTTH).

Earlier, we demonstrated the relationship between the number of combinations that can be made in an experiment and the probability of any particular outcome.

Watch Sal Khan work through a combination problem on the KHAN ACADEMY® video "9 card hands"

Not surprisingly, then, the formula for the probability of an outcome, which is the binomial function, includes the formula for combinations:

Formula 3.9: $\quad p(R) = [_nC_k]p^k q^{(n-k)}$

In coin flipping, the probability of success is 0.5, as is the probability for failure.

As an example, we can calculate the probability that exactly two heads will be the outcome of four coin flips. To begin to answer this question, we need to know how many combinations of this exist:

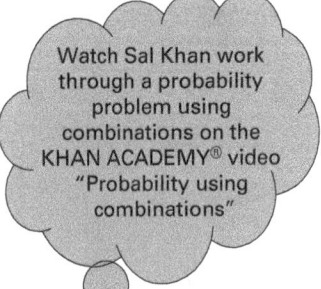

Watch Sal Khan work through a probability problem using combinations on the KHAN ACADEMY® video "Probability using combinations"

$$[_4C_2] = \frac{4!}{(2!)(4-2)!} = \frac{(4)(3)(2)(1)}{(2)(1)(2)(1)} = \frac{24}{4} = 6$$

We can insert this result into Formula 3.9:

$$p(2 \text{ heads}) = [_nC_k]p^k q^{(n-k)} = 6(0.5)^2 (0.5)^{(4-2)}$$
$$= (6)(0.25)(0.25) = 0.375$$

If we did this for every possible outcome and plotted it, we get the output displayed in Figure 3.3.

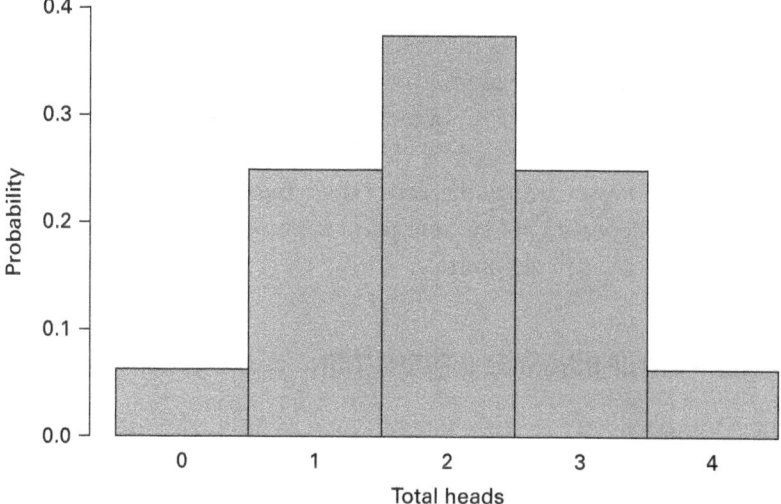

FIGURE 3.3. Probabilities of heads in four coin flips.

Notice some similarities between the histograms that we have drawn so far, which are characteristics of binomial distributions:

- They are symmetrical.
- The modal values are in the center of the distribution.

Because of these similarities, binomial distributions have two other unique features. First, the mean is defined as:

$$\text{Formula 3.10:} \quad \mu = np$$

and the standard deviation is defined as:

$$\text{Formula 3.11:} \quad \sigma = \sqrt{np(1-p)}$$

In the case of four coin flips, then:

$$\mu = np = 4(0.5) = 2$$

and

$$\sigma = \sqrt{np(1-p)} = \sqrt{4(0.5)(1-0.5)} = \sqrt{1} = 1$$

Therefore, the mean for this binomial distribution is 2 and the standard deviation is 1. This further explains what is illustrated in Figure 3.3.

In our explanation of **combinations**, notice that we were interested in the number of ways we were able to obtain a set of outcomes from a set of discrete possibilities. We were not particularly interested in the order in which the sample was chosen. There are times, however, when the order does matter and we would call these **permutations**. While we do not get into permutations any further in this book, it is something you might see in the future.

## APPLICATIONS OF THE BINOMIAL DISTRIBUTION

Throughout this book we will talk about different ways in which data can be distributed based on certain attributes of the data. Any distribution, however, is applicable only under certain conditions, and the binomial is no exception. The binomial distribution is used to describe a count variable; that is, the variable of interest must be able to be discretely counted. In the previous examples, we were able to count how many times a head was tossed, how many times a 5 was thrown from a single die, or how many times a club was dealt from a deck of cards. In any case, we need to know that the number of observations is fixed.

The binomial distribution can be applied to count variables only when all of the following conditions exist:

- Each observation is independent of the others. Notice that we did not attempt to apply the binomial when there was a dependency. In the examples provided, each flip of the coin was independent of the others.
- There are only two possible outcomes, one of which can be defined as *success* and the other as *failure*. This, however, does not mean that only two possible outcomes exist in total for an observation. For example, if we were interested in the probability of rolling a 6 when a die is thrown, we could define 6 as *success* and all other rolls as *failures*.
- The probability of success does not change for each observation. That is, every time you toss the coin, the probability of getting a head remains constant.

## COMBINATIONS AND BINOMIAL PROBABILITIES USING *R*

*R* makes it very simple to calculate the number of combinations you can make if you simply know *n* and *k*. To demonstrate, we can re-create what we did manually earlier. If, for example, we want to know how many combinations of two heads we can make in four coin flips, $n = 4$ and $k = 2$.

Enter the following into the Console:

```
> choose(4,2)
[1] 6
```

In general, the notation for the **choose()** function is **choose(n, k)**. Notice that the value for *n* is entered first inside the parentheses followed by the value for *k*.

Most of the time, however, we do not want to know how many combinations are possible, but the probability of obtaining a particular outcome. In the last example, we wanted to know what the probability was of getting two heads when four coins were flipped.

For that, we will use the **dbinom()** function where the notation is **dbinom(x, number of trials, probability of success)**.

To calculate the probability of getting exactly two heads in four coin flips, enter the following into the Console:

```
> dbinom(2, 4, 0.5)
[1] 0.375
```

*R* makes it incredibly easy to do these types of calculations, and we are glad about that, especially when the number of trials gets large. For example, instead of flipping a coin just four times, we decide to flip it 104 times. And suppose we want to know what the probability is of getting exactly 15 heads. We can do this quickly in *R*, by entering the following into the Console:

```
> dbinom(15,104,0.5)
[1] 2.352249e-14
```

Whoa! That result is so small that the answer is written in scientific notation. If you want to see this without scientific notation, simply turn that function off with one of the **options()** functions and rerun the command:

```
> options(scipen=999)
> dbinom(15,104,0.5)
[1] 0.0000000000002352249
```

The histogram illustrating 104 coin flips is shown in Figure 3.4. Notice that with success defined as exactly 15 heads, the probability of this is way in the left-hand tail of the histogram. This is reflected in the very low likelihood of such an event occurring, which we saw in the output from *R*.

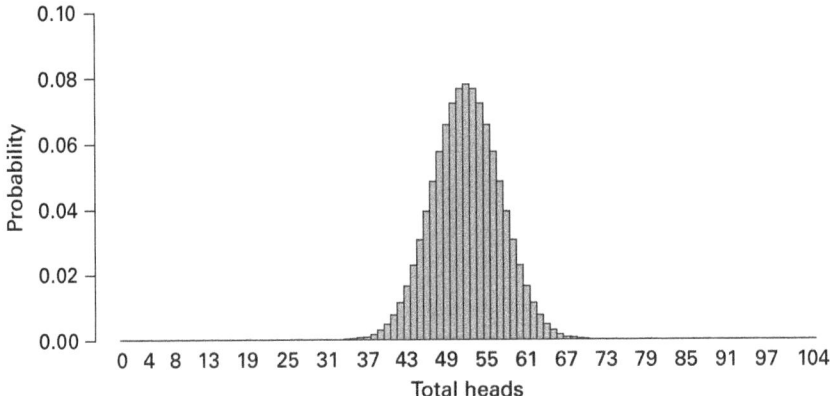

FIGURE 3.4. Binomial distribution illustrating the probability of getting heads for 104 coin flips.

Looking again at this histogram, notice that the shape of the distribution is symmetrical and the probability of getting any particular outcome may be very small, but it is never zero. Also notice that the distribution begins with zero heads and ends with 104 heads. It is finite in both directions. But also recognize that the sum of the probabilities of all 105 possible outcomes is 1, which represents the total area under the curve of the histogram.

Obviously, the binomial cannot be applied to all situations, and we will learn more about other situations and distributions in future chapters.

## CONCLUSION

Probability is the basis for many types of calculations in statistics, and it is a concept many people use frequently without even realizing it! In this chapter, we used simple experiments such as flipping coins, rolling dice, and dealing cards to illustrate some of the key concepts underlying probability.

To understand the probability of a particular outcome in an experiment, we first had to define that outcome as a *success* and everything else as a *failure*. Once that was accomplished, you learned how to calculate the probability of that success by considering whether events that made up the experiments were independent (vs. not independent) and whether a successful outcome could be considered mutually exclusive.

In this chapter, you also learned under what circumstances you could apply the binomial distribution. In cases where you could use it, you were presented with formulae that could help you calculate the probability of a successful outcome under more complex circumstances.

## EXERCISES

1. What is the probability of getting a 1 in three sequential rolls of a single six-sided die?
2. In rolling a single die six times, what is the probability of getting this exact outcome: 1, 2, 3, 4, 5, 6?
3. You draw three cards from a deck without replacing them after each draw. What is the probability that you will draw a heart followed by a diamond followed by a spade?
4. What is the probability that you will roll a 4 in a single roll of a die? What are the odds that you will roll a 4 in a single roll of a die?
5. How many possible combinations can be made by flipping a coin five times?
   a. Use Formula 3.8 to show that the sum of all possible combinations is equal to the number you calculated above using "tails" as your measure of success.
   b. Now, use Formula 3.9 to show that the sum of the probabilities for each of these outcomes is equal to 1.
   c. Calculate the mean and standard deviation for this binomial distribution.
   d. Draw the binomial distribution for this.
   e. Does your answer for part c make sense given your illustration of this binomial? Why or why not?

## *R* EXERCISES

1. A craft store has 20 kinds of beads and bracelets are made with any five. How many different combinations can be made?
2. My front door has a combination lock with five knobs. How many combinations can be made if three numbers are needed to make up the combination? What about if four numbers make up the combination?
3. What is the probability of getting 20 tails in 72 coin flips?

# 4

# THE NORMAL DISTRIBUTION AND STANDARDIZED SCORES

## INTRODUCTION

In Chapter 3, we discussed one type of distribution, the binomial. In this chapter, we will talk about the normal distribution. The normal distribution has a lot to offer those of us who are interested in statistics because of some of its unique characteristics, many of which you will learn about here.

Once you become aware of the foundations of the normal distribution, we will continue by discussing standardizing scores and the value of doing this. In this section, we will refer back to some of what you learned in Chapter 2 about cumulative frequency and percentile rank. With this knowledge in hand, you will be able to answer questions like "Is a 580 on the math section of the SAT a good score?" and "How good is it compared to other students?"

## ANOTHER TYPE OF DISTRIBUTION

The normal, or Gaussian, distribution looks similar to the binomial, which we discussed in Chapter 3, but it has a great deal more flexibility, as some of the limitations of the binomial do not apply. It is perhaps the most widely recognized class of distributions because of its easily recognizable bell-shaped curve. An example of a normal distribution is shown in Figure 4.1.

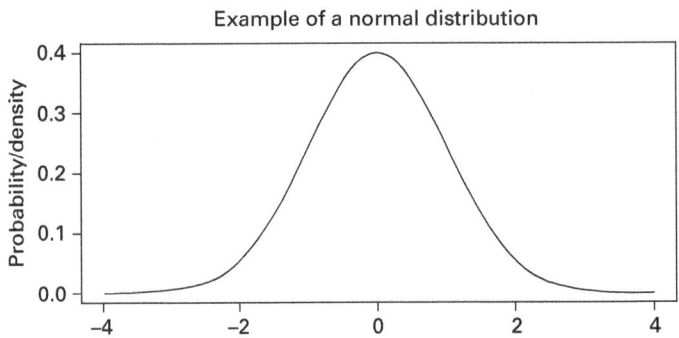

FIGURE 4.1. Normal distribution example.

Notice that in this distribution, the center point is at zero and the tails extend infinitely in both directions. Another example of a normal distribution is shown in Figure 4.2.

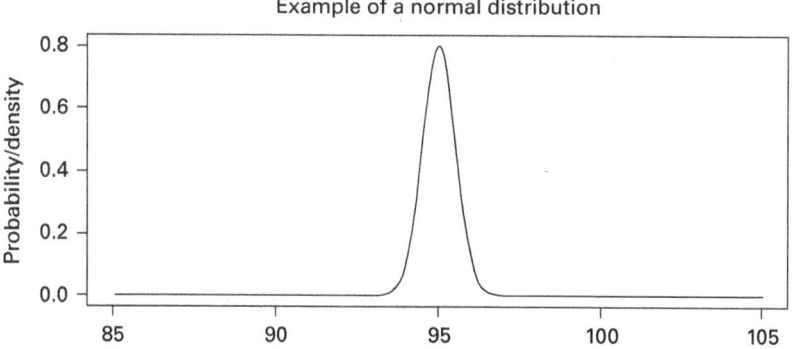

FIGURE 4.2. Another normal distribution.

While Figures 4.1 and 4.2 are both normal distributions, Figure 4.1 has a mean of 0 and a standard deviation of 1. Figure 4.2 has a mean of 95 and a standard deviation of 0.5. While their overall shapes look similar, they are, of course, not the same. Therefore, normal distributions are not a single distribution, but a class of distributions defined by the equation shown in Figure 4.3.

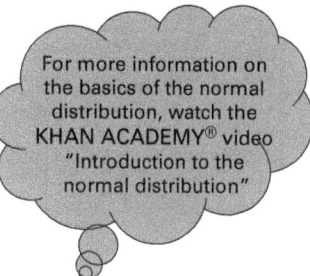

For more information on the basics of the normal distribution, watch the KHAN ACADEMY® video "Introduction to the normal distribution"

$$y = \frac{1}{\sigma\sqrt{2\pi}} e^{-\frac{(x-\mu)^2}{2\sigma^2}}$$

FIGURE 4.3. Formula for normal curve.

Luckily, you will likely have no occasion to actually plug numbers into this formula and calculate it manually. You will, however, be much more interested in some of the unique characteristics of the normal curve:

- The mean and standard deviation are used to define a normal distribution, as illustrated in the descriptions of Figures 4.1 and 4.2.
- The mean, median, and mode are all equal.
- The area under the normal distribution is symmetrical about the mean (and, therefore, the median and mode as well). It is not skewed in any way, and has that famous bell shape.
- Most interestingly, the percentage of data falling on either side of the mean is consistent and is shown in Figure 4.4.

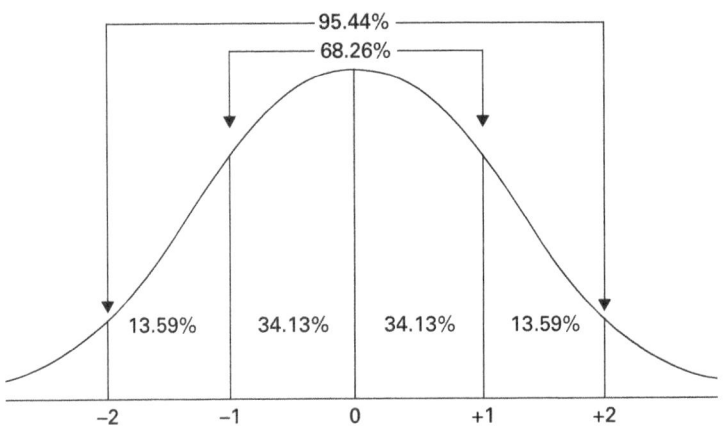

FIGURE 4.4. Normal distribution illustration with distribution of data.

For more information on how you can apply this idea, watch the KHAN ACADEMY® video "ck12.org normal distribution problems: Empirical Rule"

In Figure 4.4, the values along the x-axis denote the number of standard deviations from the mean. In this figure, we see that 68.26% of the area under the normal distribution is within one standard deviation of the mean (plus or minus) and 95.44% of the area is located within two standard deviations of the mean. Half of each of these is on either side of the mean because of the symmetrical nature of the normal distribution. Therefore, 50% of the area under the normal distribution is less than or equal to the mean and the remaining 50% of the area is equal to or greater than the mean.

To look at this more closely, consider the example of the normal distribution first described in Figure 4.2. This distribution has a mean of 95 and a standard deviation of 0.5. With this information, we know that 68.26% of the area under that distribution is between the values of 94.5 (95 − 0.5 = 94.5) and 95.5 (95 + 0.5 = 95.5). Additionally, we know that 95.44% of the area for this distribution is located between the values of 94 (95 − 2(0.5) = 94) and 96 (95 +2(0.5) = 96). The remainder of the area for that distribution, 4.56%, is located in the long tails extending out from the high peak.

If we look at these characteristics of the normal distribution, we know that the area under the curve is equal to 1, or 100% of the area.

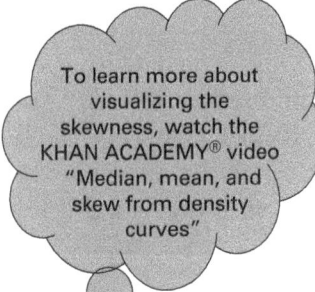

To learn more about visualizing the skewness, watch the KHAN ACADEMY® video "Median, mean, and skew from density curves"

### More on the Shape of the Normal Distribution—Skewness and Kurtosis

**Skewness** describes how symmetrical a distribution is, while **kurtosis** describes how peaked or flat a distribution is. Figure 4.5 illustrates both positively and negatively skewed distributions.

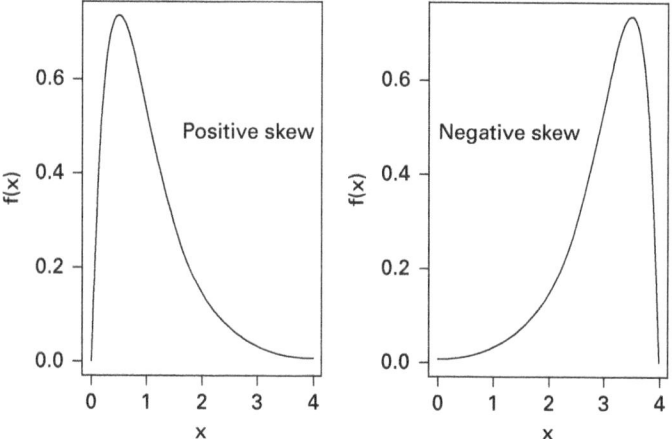

FIGURE 4.5. Distributions with skewness.

Sometimes it is helpful to quantify the amount of skew a distribution has. If the mean is greater than the median, the distribution will be skewed to the right and be described as a positive value. Distributions skewed to the left will produce a negative value. A perfectly normal distribution will have a skew of 0.

Figure 4.6 shows distributions with leptokurtosis and platykurtosis. Notice that the leptokurtotic distribution is more highly peaked than the normal distribution, while the platykurtotic distribution is flatter.

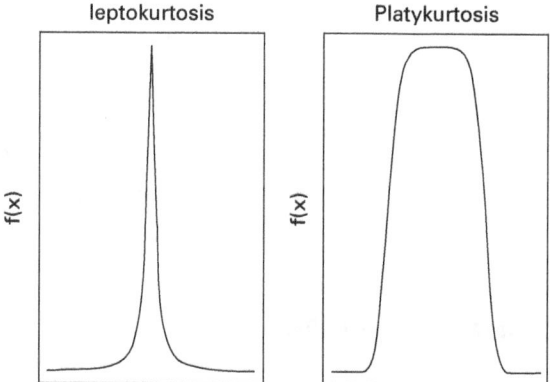

FIGURE 4.6. Kurtosis in distributions.

You can use the **describe()** function illustrated in the previous chapter to view the skewness and kurtosis of any variable. Kurtosis values of 0 denote a normal distribution, those with positive values denote more leptokurtosis, and those with negative values denote more platykurtosis.

## WHY IS THE NORMAL DISTRIBUTION SO SPECIAL?

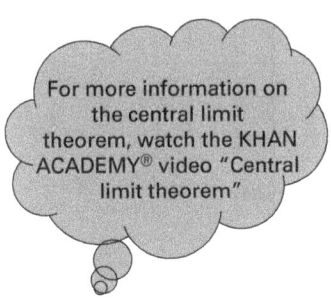

For more information on the central limit theorem, watch the KHAN ACADEMY® video "Central limit theorem"

The normal distribution is fundamental to the field of statistics. This is due to the central limit theorem, which states that if repeated samples are taken from any distribution, the shape of that distribution, the sampling distribution, approaches normality as the number of samples taken from the original distribution increases. Additionally, the law of large numbers tells us that the sample mean of the sampling distribution will equal the mean of the population (i.e., $\bar{x} = \mu$), and the standard deviation of the sample (i.e., the standard error of the mean or standard error, for short) will be $\sigma_{\bar{x}} = \sqrt{\dfrac{\sigma^2}{n}}$.

What is noteworthy about this is that these features are maintained regardless of the shape of the original distribution. As a result, we can apply what we know about the normal distribution to any sample provided that the sample itself is sufficiently large. This leads us to the question, "What is a sufficiently large sample?"

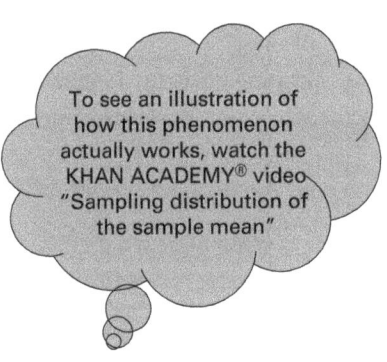

To see an illustration of how this phenomenon actually works, watch the KHAN ACADEMY® video "Sampling distribution of the sample mean"

While you may think that "sufficiently large" could mean sample sizes that are huge, that is not necessarily the case. Statisticians have agreed, in fact, that this number is actually fairly small, with sample sizes ranging somewhere between 15 and 40 (Allen, 1990; Casella & Berger, 1990; Cherry, 1998; Moore & McCabe, 1989).

When this is the case, we can make use of the central limit theorem to make inferences about the relationships between variables since these statistical tests are based on samples and not a population.

## THE STANDARD NORMAL DISTRIBUTION

In Figure 4.1, you saw a special version of the normal distribution in which $\mu = 0$ and $\sigma = 1$. This is called the standard normal distribution or the *z*-distribution. When we can assume that our data are normally distributed and we obtain a value for a variable, we can apply what we know about the normal distribution to understand what percentile of total values falls above or below that value by standardizing the original value.

z-Scores define the distance of a raw score from the mean in terms of standard deviations using the following formula:

$$\text{Formula 4.1:} \quad z = \frac{X - \mu}{\sigma}$$

z-Scores can have both positive and negative values. Positive values fall to the right of the mean in the standard normal deviation and negative scores fall to the left.

To illustrate this application, we can consider the example of Joe College, who just took his semester midterms for three classes. He received a 63 on his statistics test, an 85 on his psychology test, and an 85 on his sociology test. He had studied long and hard for his statistics exam, so imagine his disappointment when he found his grade online! He, of course, felt much better about his sociology and psychology exams.

During the class after the midterm was given, Joe's statistics professor provided the mean and standard deviation for the test for the entire class. Joe, being the diligent student that he is, asked his other professors for the mean and standard deviations for his other exams. This is what he found:

| | $x$ | $\mu$ | $\sigma$ |
|---|---|---|---|
| Statistics | 63 | 60 | 6 |
| Psychology | 85 | 90 | 5 |
| Sociology | 85 | 85 | 4 |

Joe's score is in the first column labeled $x$ and the class means and standard deviations are in the following two columns. When we compute a z-score for Joe's grades, he can then accurately compare how he did to the rest of the class.

Begin by computing a z-score for Joe's statistics exam:

$$z = \frac{X - \mu}{\sigma} = \frac{63 - 60}{6} = \frac{3}{6} = 0.5$$

This result shows that Joe's score on the statistics test was half a standard deviation above the mean for the class, and is illustrated in Figure 4.7.

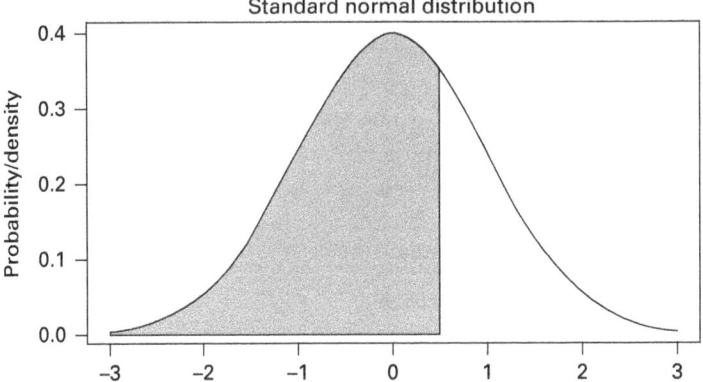

FIGURE 4.7. Area under normal curve with a z-score of 0.5.

Because we are interested in the people who did worse than Joe (i.e., they had scores lower than his), we shade the area to the left of the z-score of interest.

Now we can use the z-distribution table, Table 1 in Appendix F, to determine how he did compared to others in the class.

### USING THE Z-DISTRIBUTION TABLE

To bring understanding to Joe's score, you will need to refer to the z-distribution table. To utilize it, simply locate the score you computed to the tenths place in the far left column and then go across the top of the table for the hundredths place. Since 0.5 is the same as 0.50, simply locate that value on the table where these two values intersect, which is 0.6915. This means that Joe's score was better than 69.15% of the class. Not bad considering he got a 60%!

We can compute z-scores for Joe's other exam scores as well. Let's continue with psychology:

$$z = \frac{X - \mu}{\sigma} = \frac{85 - 90}{5} = \frac{-5}{5} = -1$$

To visualize this, we can draw the standard normal curve and shade in the area of interest; in this case, this is the area to the left of −1, as illustrated in Figure 4.8.

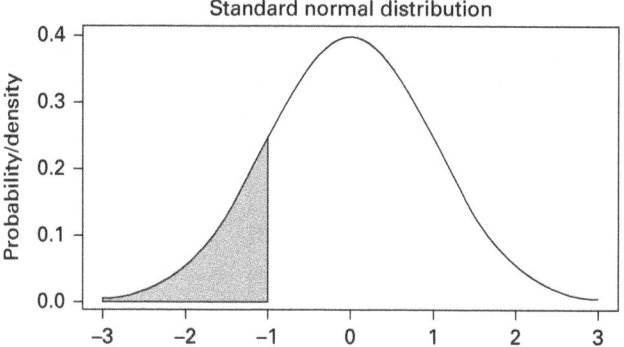

FIGURE 4.8. Area under normal curve with a z-score of –1.0.

This indicates that Joe's score was one standard deviation lower than the rest of his class. This is illustrated in the seemingly small amount of area shaded in Figure 4.8. Despite a seemingly high grade, Joe did not do well compared to his classmates.

For more examples of z-score problems, watch the KHAN ACADEMY® video "ck12.org normal distribution problems: z-score" and "ck12.org: More empirical rules and z-score practice"

The z-score for Joe's sociology exam can be calculated, as follows:

$$z = \frac{X - \mu}{\sigma} = \frac{85 - 85}{4} = \frac{0}{4} = 0$$

With a z-score of zero, we know that Joe's grade is exactly at the mean of the class, as we saw from the table. This is illustrated in Figure 4.9.

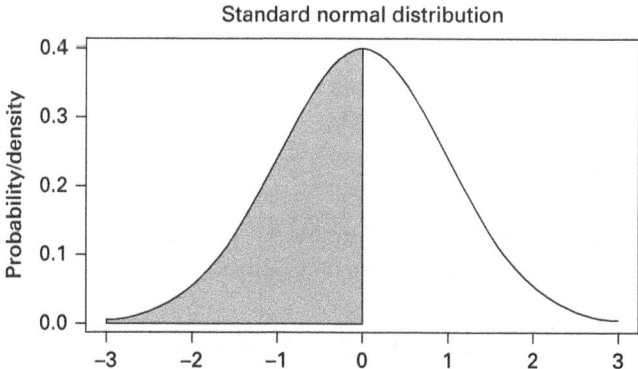

FIGURE 4.9. Area under normal curve with a z-score of 0.

Looking at the $z$-score table, you can see that Joe scored in the 50th percentile, with half the class doing better than him and half doing worse. Notice that Figure 4.9 illustrates this clearly with half of the normal distribution being shaded and half not.

Now we can expand the table we created previously to really compare Joe's grades:

|  | $x$ | $\mu$ | $\sigma$ | $z$ | % rank |
|---|---|---|---|---|---|
| Statistics | 63 | 60 | 6 | 0.5 | 69.15 |
| Psychology | 85 | 90 | 5 | −1 | 15.87 |
| Sociology | 85 | 85 | 4 | 0 | 50 |

Despite Joe's seemingly low grade in statistics, he actually did better in this class relative to his peers than in his other two classes. While Joe received the exact same grade for his tests in psychology and sociology, compared to his classmates, he did average in sociology, but not very well in psychology. How would you advise Joe to use his study time for finals?

## OTHER TYPES OF PROBLEMS USING Z-SCORES

The previous example is a very typical application of $z$-scores, but there are two other types of problems that you might see and several other ways to interpret values shown on $z$-distribution tables.

In the previous example, you were given a raw score (Joe's grades) and were asked to determine a percentile based on that score. A variation of this is the opposite: you are given the mean of a sample, and you are asked to find a raw score given a $z$-score or, in other cases, a percentile.

To illustrate this, consider IQ scores with a known $\mu = 100$ and $\sigma = 10$. What IQ score would someone need to have to be considered very smart, say, in the 95th percentile? To begin, you need to obtain a $z$-score for the 95th percentile. To get this value, look within the $z$-distribution table itself to see where the 95th percentile lies. The closest values are 0.9495 (for a $z$-score of 1.64) and 0.9505 (for a $z$-score of 1.65), so we can get as close as possible by going right between those with a $z$-score of 1.645.

To solve this problem, rearrange Formula 4.1 to isolate $x$, which is the raw score for which we need to solve. The result is Formula 4.2:

$$\text{Formula 4.2}: \quad x = z \cdot \sigma + \mu$$

Now, simply enter the known values to solve for $x$:

$$x = z \cdot \sigma + \mu = 1.645 \cdot 10 + 100 = 16.45 + 100 = 116.45$$

Someone with an IQ of 117 (since these scores are measured in whole numbers) scores in the top 5% of IQ scores, since his or her score is better than 95% of others.

Another type of problem asks you for a range of scores, and in these cases, you will need to think critically about the information you are obtaining from the z-distribution tables to answer these questions accurately.

Here's an example: "What percentage of people have IQs between 75 and 90?" Unlike our previous examples, we do not simply want the values to the left of a single value. Instead, we want some shaded area of the distribution since we are now looking at a range of scores and not simply a single value. This is illustrated in Figure 4.10.

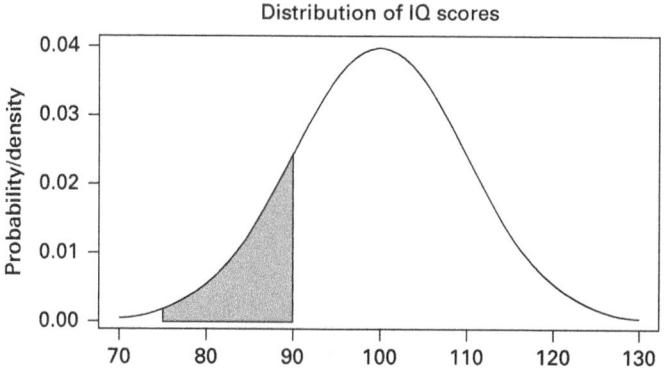

FIGURE 4.10. Percentage of people with IQs between 75 and 90.

This problem is different than the others we have seen because we need to eliminate scores that are less than 75, whereas in previous examples, we would want the entire area under the curve that is less than, or to the left of, some critical value. To tackle this type of problem, then, we will have to include the entire area between 75 and 90 but eliminate the area that is less than 75 by subtracting it.

To begin, calculate the z-score for the raw score of 90:

$$z = \frac{X - \mu}{\sigma} = \frac{90 - 100}{10} = \frac{-10}{10} = -1$$

Using the z-distribution table, the area under the normal curve that is less than a z-score of −1 is 0.1587.

Now calculate the z-score for an IQ of 75:

$$z = \frac{X - \mu}{\sigma} = \frac{75 - 100}{10} = \frac{-25}{10} = -2.5$$

With this knowledge, we only want to know the area of the standard normal distribution with these qualities: $-2.5 \leq z \leq -1$, as illustrated in Figure 4.11.

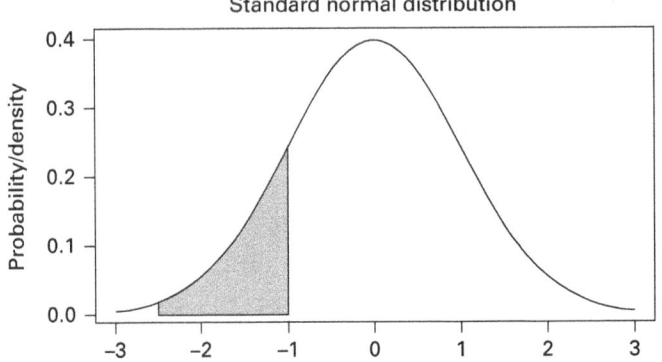

FIGURE 4.11. Illustration of standard normal distribution with -2.5 ≤ *z* ≤ -1.

Referencing the $z$-distribution table, notice that the corresponding area is 0.0062.

To determine the area between IQ scores of 75 and 90 and, correspondingly, $2.5 \leq z \leq -1$, simply subtract:

area = 0.1587 − 0.0062 = 0.1525

To answer the original problem of "what percentage of people have IQs between 75 and 90?" simply change the resulting answer from a proportion into a percentage: 15.25%.

Other types of problems that also require some additional subtraction are questions that ask about scores greater than some critical value. This is because it entails looking at the area to the right of the critical value in Table 1, whereas that table gives us the area to the left of a critical value. Here's an example: "What percentage of people have IQs greater than 90?"

In the previous problem, we already calculated the $z$-score for an IQ of 90 and obtained the area under the standard normal curve that corresponds to a $z$-score of −1 or less: 0.1525. Now, however, we are interested in the area above this, as illustrated in Figure 4.12.

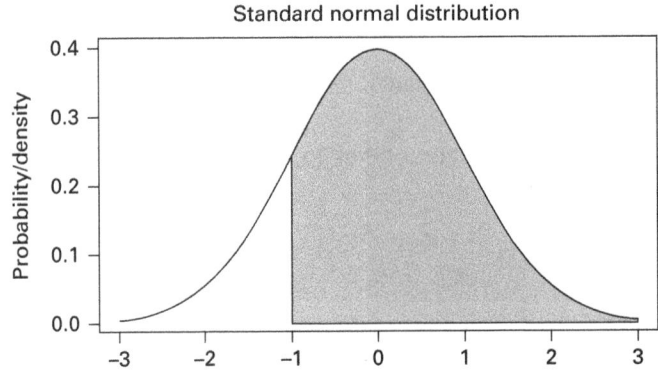

FIGURE 4.12. Looking for the area under the normal distribution with IQs greater than 90.

Since the entire area of the standard normal curve is 1, to answer this question, simply subtract 0.1525 from 1:

$$1 - 0.1525 = 0.8475 \text{ or } 84.75\%$$

In general, we recommend you address these types of problems in a systematic manner:

1. Determine what information you have and what information you need. Based on this, choose to utilize either Formula 4.1 or 4.2. You will want to use the formula in which your unknown value is isolated on one side of the equation.
2. Do your calculations and use Table 1, not necessarily in that order.
3. Draw a standard normal distribution shading the area of interest based on the question.
4. Determine whether you can simply use the values in Table 1 or whether you need to do some additional math to answer your original question. Then do it!

## CONCLUSION

In this chapter, we looked at the normal distribution, a foundational concept in the study of statistics. In cases where the number of observations is sufficiently large, we can use characteristics of the normal distribution to draw certain conclusions. These will be discussed further in later chapters.

The standard normal distribution is a special case of the normal distribution in which the $\mu = 0$ and $\sigma = 1$. This distribution is also called the $z$-distribution, and $z$-scores can be calculated to standardize scores when the mean and standard deviation of the population are known. We also illustrated how $z$-scores and $z$-score tables could be used to determine percentiles in a variety of circumstances.

## EXERCISES

For each of the following problems, assume that SAT scores are normally distributed with $\mu = 1,000$ and $\sigma = 100$.

1. Sally's combined score on the SAT was 1220. What percentage of students did worse than her? What percentile is she in?
2. Justin is shooting to get into an Ivy League school and knows his chances are better if his SAT scores are in the top 0.1% of college applicants. What score does he need to achieve this?

3. The first time Jasmine took the SAT, her total score was 990. She studied hard and raised her score to 1110. How much did she move up in terms of percentile?
4. Ricardo told his friends that he scored better on the SAT than 88.5% of other students. What was his score?
5. Sarah's score was exactly the average SAT score. By how many points must she raise her score if she wants to be in the 75th percentile?

# 5

# INTRODUCTION TO INFERENTIAL STATISTICS

## INTRODUCTION

So far, we have laid the groundwork for more complex concepts in statistics by introducing topics such as displaying and describing data. In Chapters 3 and 4, we complicated this by talking about probability and distributions. Together, this enabled us to begin to make comparisons. For example, we were able to take an IQ score and, by drawing on what we know about the normal distribution, determine how one score compared to others.

Very often, however, we are not interested in single scores or single observations. More commonly, we are provided with a certain amount of information about, say, a group of people (i.e., a sample), and we need to draw inferences about some larger population from the sample. Inferential statistics helps us do just that.

In this chapter, we will introduce topics necessary for delving into inferential statistics. We will expand upon our discussion of the central limit theorem and will introduce the notion of hypothesis testing. From there, we will talk about statistical significance and then decide where to "draw the line" on what is considered such an unusual event that we should sit up and take notice. Later in the chapter, we will talk about the limits to hypothesis testing. Finally, we will discuss how to select tests of statistical significance.

## INFERENTIAL STATISTICS

Very frequently, we may want information about some population, but we do not have access to everyone or everything that makes up that population or it is inefficient to study the entire population. In those cases, we might obtain data from a sample and then, using statistical techniques, determine whether or not we can draw some conclusions about the larger population.

Sampling, the act of selecting elements from a population, requires a great deal of skill, and entire texts are written about it. We cannot overstate the importance of strong sampling methods; if a sample is not representative of the population of interest, conclusions that you may draw due to your statistical analysis are likely to be erroneous.

In general, there are two types of samples, probability and nonprobability samples. In a **probability sample**, every element of the population has an equal and known chance of being included in the sample. The techniques utilized to create a probability sample always employ some type of random selection. For example, if our population of interest was college students in Los Angeles, to draw a true probability sample, we would need to obtain the names of all college students in Los Angeles and use some sort of technique to randomly select people to be in our sample. **Nonprobability samples**, on the other hand, do not require that every element of the population has an equal and known chance of being sampled. One of the most common types of sampling done in the social sciences is a convenience sample, which is a nonprobability sample based on some type of convenient feature. For example, if our population of interest was college students in Los Angeles, we may try to sample students who are entering or leaving the Student Union at UCLA. While each of those individuals may be a student, they probably do not represent all students in Los Angeles because they all attend one school and were visiting a building where students congregate.

While we do not go into the details of sampling this text, we strongly recommend you refer to some of the excellent texts listed in Appendix A prior to conducting your own research. Throughout the remainder of this text, when we discuss samples, we assume that your sample has been randomly selected and is representative of a larger population.

In general, we can answer two types of questions using inferential statistics, both of which are related to how similar a sample statistic is to a population parameter. The most commonly asked questions include:

- How *distant* is my sample mean from the population mean?
- How *close* is my sample mean to the population mean?

When our questions are formed around distance, we get involved with hypothesis testing and look at statistical significance and power. When our questions are formed around closeness of a statistic to a parameter, we frequently look at confidence intervals. These are all complicated topics, and we get into much more detail about each of these in subsequent chapters.

## FORMING HYPOTHESES

By making use of the central limit theorem, as discussed in Chapter 4, we can, in many instances, test hypotheses; however, to test hypotheses, we first need to form them.

In general, there are two types of hypotheses, directional and nondirectional. An example of a directional hypothesis would be "Those receiving Intervention A will have higher mean levels of self-esteem than those receiving treatment as usual (TAU)." To put it more simply:

$$H_1: \mu_A > \mu_{TAU}$$

Here is another example, "Students receiving individual tutoring (T) will have lower average GPAs than students receiving group tutoring (G)." Again, to state this hypothesis mathematically, we could simply write:

$$H_1: \mu_T < \mu_G$$

A nondirectional hypothesis differs in that it only suggests that a relationship exists between two variables but does not predict the nature of the relationship. For example, "Senior citizens who attend programs at the local community center (P) will differ in the average number of social contacts they have each week from seniors who attend programs at the local church (C)." Written mathematically:

$$H_1: \mu_P \neq \mu_C$$

Notice here that we are not suggesting that participation in the community center program is related to more or less social contacts than participation in church programming—only that it is different.

Here is another example: "Patients who take Drug A for schizophrenia have a different mean number of auditory hallucinations than patients who take Drug B." Or:

$$H_1: \mu_A \neq \mu_B$$

In each of these examples, notice that our stated hypothesis is denoted by $H_1$. This can also be written as $H_A$ to denote the alternate hypothesis.

The opposite of any hypothesis that you may form is the null hypothesis. Simply stated, the null hypothesis is the hypothesis of no difference or no relationship and covers all other possibilities than those expressed in the alternate. For example, the alternate hypothesis constructed for self-esteem in the intervention versus treatment-as-usual group mentioned earlier was directional: $H_1: \mu_A > \mu_{TAU}$. Here, the null hypothesis would be:

$$H_0: \mu_A < \mu_{TAU},$$

which states that those in the treatment-as-usual group have the same or higher levels of self-esteem than those participating in Intervention A.

When constructing hypotheses, notice that the null and alternate must be mutually exclusive and exhaustive. Between the two of them, all possible eventualities should be covered.

Looking at a nondirectional hypothesis, the alternate hypothesis in the schizophrenia study was $H_1 : \mu_A \neq \mu_B$. The nondirectional null, then, would be $H_0 : \mu_A \neq \mu_B$, which states that there is no difference in the mean number of auditory hallucinations between those taking Drug A and Drug B. They are equal. Another common way to state this is $H_0 : \mu_A - \mu_B$. That is, the difference in the mean number of hallucinations between those taking Drug A and Drug B is zero.

Notice that in writing all these hypotheses, we state what we think is happening in terms of the population; after all, that is what we are interested in. Despite this, in most cases, we only have access to a sample, so we will have to use statistical inferences to test our hypotheses. As we begin to test hypotheses, we begin with the assumption that the null is true. If we are able to reject the null hypothesis, we can accept the alternate since it is the opposite of the null. If we are unable to reject the null, we continue to accept it and reject the alternate since both statements cannot be true, as each is mutually exclusive.

This provides quite a challenge since we are unable to observe what is happening in the population, but we can observe what is happening in a sample. When we use inferential statistics, we may be able to draw conclusions about the population from the sample.

## YOU MAY BE WRONG, BUT YOU MAY BE RIGHT

Like the Billy Joel song, the conclusions we draw based on our hypothesis testing may be wrong, but they may be right; however, there is really no way for us to know for sure.

We could observe a change in the sample or we could notice differences between groups in the sample that are not actually occurring in the population. What we observe may have happened by chance. Alternatively, our sample may not reflect differences when they in fact exist in the population. Table 5.1 illustrates the various possibilities.

TABLE 5.1. Possible Scenarios Between What Is Observed and What Exists in a Population

| | | Population | |
| --- | --- | --- | --- |
| | | Relationship | No Relationship |
| Sample | Observe a Relationship | CORRECT! | Type I Error |
| | Observe No Relationship | Type II Error | CORRECT! |

If we are correct in our assessment, then all is right with the world, but if we are wrong, this could lead to erroneous decision making. While there is no way to tell what is happing in a population without observing the entire population, we can calculate our chances of drawing an incorrect conclusion.

When testing hypotheses, we are testing for **Type I error**, which is that observed differences in our sample do not actually exist in the population. This can be thought of as a "false positive." Erroneous decision making due to Type I error can have serious consequences. For example, what if we thought a treatment for schizophrenia was effective when in fact it was harmful? Similarly, what if we thought that an educational intervention for autism improved children's communication skills when it actually made them worse? Clearly, making a Type I error can have deleterious effects.

For more information on Type I error, watch the KHAN ACADEMY® video "Type 1 errors"

**Type II error**, which is that we do not see observed differences in our sample that actually exist in the population, can be thought of as a "false negative." In the example earlier, a Type II error would be thinking that a new intervention for schizophrenia was not effective because we did not observe changes in our sample; however, the intervention is in fact effective. Because of some type of problem with our sample, we just did not observe those changes.

You should be aware, then, that there is an inverse relationship between Type I and Type II error. As the chances of making a Type I error increase, the chances of making a Type II error decrease, and vice versa. Researchers and statisticians, then, must do their best to find a middle ground in balancing these. In large part, this balance is struck by considering what could be an acceptable chance of making a Type I error while simultaneously understanding the risk of making a Type II error, which can be manipulated by adjusting the sample size (i.e., as a sample size approaches the size of the population, it is unlikely that we would not observe something in the sample that exists in the population). We will address this Type II issue later in the book.

When we test hypotheses, we are looking at what our chances are of making a Type I error. In general, for a finding to be considered statistically significant, the chance of making a Type I error should be low and is set by a **criterion of significance**, which is a threshold above which we cannot reject the null hypothesis since the chance of making a Type I error is considered too high to be acceptable.

The criterion of significance is denoted by the Greek letter alpha ($\alpha$). In the behavioral and social sciences, the most commonly accepted cut-off for a criterion of significance is $\alpha = 0.05$. That is, an acceptable rate for making a Type I error is 5% or less. This means that out of every 100 experiments, we are willing to accept that we are wrong in assuming that observed differences in the sample are reflected in the population five times. Any test of Type I error above this rate would be considered unacceptably high, and we would have to conclude that any relationship we observe in our sample is not actually reflected in the population. Because of this, we would be unable to reject our null hypothesis, whatever that hypothesis may be.

## TESTS OF TYPE I ERROR LOOK FOR UNUSUAL EVENTS

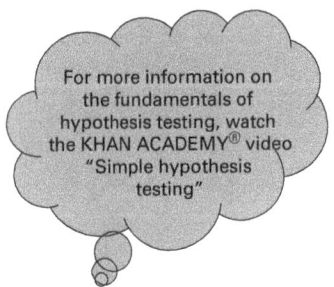

For more information on the fundamentals of hypothesis testing, watch the KHAN ACADEMY® video "Simple hypothesis testing"

If a study is well designed, observations that vary from what is expected significantly lead us to reject the null hypothesis and accept an alternate. As an example, let us think about flipping a coin like we did in Chapter 3. The experiment we are going to undertake is flipping a coin to determine if the coin is fair. In terms of hypothesis testing:

$H_0$: $f_h = f_t$ (i.e., the frequency of getting heads and tails is the same, and the coin is fair)

$H_1$: $f_h \neq f_t$ (i.e., the frequency of getting heads and tails is not the same, and the coin is not fair)

Since we will follow the conventions of the behavioral and social sciences, we will set $\alpha = 0.05$.

When we start flipping the coin, heads come up three times in a row. Do we think the coin is fair?

$$p(3 \text{ heads come up}) = (0.5)^3 = 0.125$$

A 12.5% chance of getting heads three times in a row is pretty small, but it is above our criterion. Still, this seems somewhat unusual. Let's flip three more times and determine what the probability would be of getting heads six times in a row:

$$p(6 \text{ heads come up}) = (0.5)^6 = 0.0156$$

There would be a very low chance of getting six heads in a row, and if we did, with $\alpha$ set at 0.05, we could reject the null hypothesis and accept the alternate. We could assume that the six-heads-in-a-row coin was not a fair coin. Still, it is possible to get six heads in a row with a fair coin (in fact, the probability of that happening is 1.56%), and if we reject the null hypothesis and our coin was actually a fair coin, we would be making a Type I error. Still, with such a low probability of this happening, the decision to consider the coin not fair is not unreasonable since you already set the criterion of significance at 5%.

As you continue to learn about hypothesis testing through the remainder of this book, you will want to do what we did earlier as you tackle any of the problems presented:

1. State the null and alternate hypotheses—we recommend actually writing these down as you begin working because you will need to refer back to them later.
2. Set your $\alpha$ level. While it is likely to remain at 0.05, there are conditions under which it might be different.
3. Choose a statistical test. This will dictate which test statistic you use. Once you do this, you can look up the critical value on the tables in Appendix F based on the $\alpha$-level set.
4. Calculate the actual test statistic that applies to your data. Because of the wonders of statistical software, when you use $R$ to conduct a statistical test, you will obtain a test statistic, but you will also obtain a $p$-value, which is the probability of obtaining a result as extreme as or more extreme than the observed results if the null hypothesis is true. If you use statistical software and obtain a $p$-value, you will not need to use Appendix F.
5. Make a decision about your hypothesis by comparing your calculated test statistic to your critical value (i.e., your predetermined cut-off value). If your calculated test statistic is greater than or equal to the critical value, then your observations are considered such unusual events that you can reject the null hypothesis and accept the alternate. If the calculated test statistic is less than the critical value, then your observations are not considered unusual, and you would be unable to reject the null. When using $R$ to compute $p$-values, you will simply compare the calculated $p$-value you obtain when you view the results of a statistical test with $\alpha$. If the calculated $p$-value is above $\alpha$, you will not be able to reject $H_0$. If it is less than or equal to $\alpha$, you will be able to reject $H_0$ and accept $H_1$. Stated mathematically, if $p > \alpha$, accept $H_0$; otherwise (i.e., if $p \leq \alpha$), reject $H_0$ and accept $H_1$.

## WHAT INFLUENCES STATISTICAL SIGNIFICANCE?

Later on, we will learn about the different tests of Type I error and the conditions under which you would want to use a particular test, but in general, there are three factors that influence the magnitude of $p$-values and, therefore, the chances of making a Type I error:

For more information on the relationship between hypothesis testing and $p$-values, watch the KHAN ACADEMY® video "Hypothesis testing and $p$-values"

1. Observed differences—the larger the differences between what is **expected** and what is observed, the more likely it is that

what is observed did not happen by chance and likely exists in the population. This is what was illustrated in the coin-flipping example earlier. If you are comparing two groups, for instance, those participating in Intervention A and those in the TAU group, the expectation in a null hypothesis is that there is no difference between the groups. In any event, the larger a difference that is observed between the groups, then, the more likely it is that your findings are statistically significant (i.e., the calculated *p*-value is lower).

2. Sample size—in general, as sample size increases, the chances that findings are statistically significant increases. This is logical because as a sample size continues to increase, it eventually approaches the size of the population, and the closer you get to observing everyone in a population, the more representative the sample is likely to become.

3. Variation—most of the time, the more variation that is observed in the sample, the less likely it is that findings from a test of Type I error will be statistically significant. This is because when observations vary widely, it is hard to ascertain that what is being observed is "typical" for the population. On the flip side, lower levels of variation typically increase the likelihood that findings are statistically significant. In some cases, however, very little variation could indicate that the sample is not representative of the population, and the sampling methods used should be considered.

It is important to consider what contributes to statistical significance because very often the decision to accept or reject a null hypothesis is based on a somewhat arbitrary threshold, $\alpha$, and this decision is a dichotomous one—either accept or reject the null hypothesis. There is no such thing as findings being a "little significant" in the same way that no one can be a "little pregnant." Findings of tests of Type I error are either significant or not. Knowing what contributes to *p*-values can help us understand findings in context.

For example, assume that the experiment testing Intervention A versus TAU was a pilot study and only 30 people participated, with 15 people in the sample being randomly assigned to participate in Intervention A and the other 15 people receiving treatment as usual. Everyone in both groups completed the Rosenberg Self-Esteem Scale at the conclusion of the study.

In this case, we can state our hypotheses as $H_0: \mu_A = \mu_B$ and $H_A: \mu_A \neq \mu_B$, and we will set $\alpha = 0.05$.

After the experiment, the mean self-esteem score for the intervention group was 35 with a standard deviation of 6, while the mean value for the treatment-as-usual group was 32 with a standard deviation of 4. The calculated *p*-value for the test of Type I error was 0.1183. While those in the intervention group had higher levels of self-esteem, we could not reject the null hypothesis because $p > 0.05$. This was the case even though scores in the intervention group were 3 points higher than in the TAU group; however, because of greater variation in the intervention group and a small sample size, it might be prudent to replicate this experiment with a larger sample. We might not want to prematurely call Intervention A ineffective because our findings were not statistically significant. If we did that and Intervention A was effective in reality, our caution at not rejecting the null hypothesis actually caused us to make a Type II error, which is the equivalent of a "false negative."

## A WORD OF CAUTION AND CRITICISMS OF HYPOTHESIS TESTING

While hypothesis testing has its place in the field of statistics, it is by no means definitive, and incorrect decisions can be made. One of the major criticisms in the use of statistics in the behavioral and social sciences is the overreliance of hypothesis testing in decision making. After all, once a study is completed and all the numbers have been crunched, we, as researchers, have to determine what our findings mean and we must decide what to do with those findings. Research findings often have real implications for developing and funding programs, treating clients, or teaching students. Therefore, it is incumbent upon us that we are careful in our analysis and interpretation of findings.

Some of the more commonly mentioned criticisms of hypothesis testing are as follows (Franco, Malhotra, & Simonovits, 2014; Murphy, Myors, & Wolach, 2014; Schneider, 2013):

- The null hypothesis is almost never true. That is, it is uncommon that there are actually NO differences between groups. Usually there is some difference that may result in possibly a small effect (either positive or negative) that may have clinical or practical implications in real-life settings.
- Statistical power is closely related to findings from hypothesis testing. The power of statistical tests is related to Type II error and describes the same type of relationship that was discussed previously between tests of Type I error (i.e., hypothesis testing) and $p$-values. Insufficient power, which is often related to sample sizes being too small, may lead to erroneous findings in hypothesis testing. Earlier, we provided you an example of a study with a very small sample size that might have resulted in not rejecting the null when the alternate was, in fact, true. However, the opposite can occur. When samples are very large, calculated $p$-values can result in significant findings even when the magnitude of observed change is very small and may not have any practical meaning in real life.
- Criteria of significance are somewhat arbitrary. While it is true that $\alpha$ is traditionally set at 0.05 in the social or behavioral sciences, other $\alpha$ levels are found in the literature including 0.01, 0.03, and 0.10. This, combined with a publication bias of only publishing findings that are deemed statistically significant, often leaves important studies unpublished and results inaccessible to research consumers.

- Overreliance on statistical tests leads to "all or nothing" thinking. As discussed previously, findings are either statistically significant or are not. This often leads to researchers basing important decisions on calculations of a single $p$-value when, perhaps, more holistic approaches would be appropriate. In general, findings that are not deemed statistically significant are rejected and those that are significant are accepted.

While there are criticisms for using tests of Type I error in statistical analysis, we teach hypothesis testing in detail in this text because it is the prevailing norm in most research settings, and it can be used in conjunction with additional measures, such as effect sizes and confidence intervals, that may provide a better, more nuanced analysis of research data.

Effect sizes are often used to describe magnitude of change or difference between groups. For instance, in the example of Intervention A compared to TAU, we could calculate an effect based on Cohen's $d$, the most commonly used and reported measure of effect size. While you will learn to calculate Cohen's $d$ later in this text, we can determine that for those utilizing Intervention A, there was a medium effect compared to TAU. This finding has a qualitative component attached to it, as these calculations often are best at describing the degree of change that is observed. Because of this, effect size measures are often used to describe clinical or practical significance as opposed to statistical significance. You will learn more about calculating and interpreting effect sizes in subsequent chapters.

Another measure that is frequently used in conjunction with hypothesis testing and effect size describes, with a certain degree of confidence, a range where we believe a population parameter lies. These are called **confidence intervals**. Again, you will learn more about how to calculate and interpret confidence intervals later in this book.

You should be aware that there is a school of thought among statisticians and researchers that effect sizes bounded by confidence intervals should actually replace hypothesis testing altogether. While we teach you how to calculate and interpret both effect sizes and confidence intervals in this book, we also want to emphasize that, at the time of this printing, the prevailing norm in most published social science research is to primarily rely on traditional hypothesis testing.

## GETTING STARTED WITH HYPOTHESIS TESTING

Earlier in this chapter, we mentioned that there are many different tests of Type I error, and here we will get you thinking about which test to use given the nature of your data. In general, there are two forms of statistical tests: parametric and nonparametric.

Parametric tests are favored if the assumptions for using them are met. This is because, given certain assumptions, we can apply a known sampling distribution, such as the normal distribution. This makes the results of these statistical tests more powerful in deciding to reject the hypothesis when the null is false. You will want to consider using a nonparametric statistical test if the following are true:

- You believe your observations are not independent of one another. For example, in survey research, it is generally assumed that Fernando's responses are in no way related to Jennifer's. If this is not the case, you will want to consider using a nonparametric test. Some nonparametric tests, however, do require that observations be independent, so this is not a blanket rule. Still, in these cases, the option to use a nonparametric test should be considered.
- Your data contain a lot of **outliers**. Outliers are those observations whose values vary widely from others. For example, a sample of five students were asked to provide the number of hours they worked in the last week:

| Student | Hours Worked |
|---------|--------------|
| 1       | 10           |
| 2       | 15           |
| 3       | 7            |
| 4       | 9            |
| 5       | 52           |

Looking at this data, we see that four of the five students worked a similar number of hours, but Student 5 worked much more than everyone else, and he would be considered an outlier.

In a small sample, it is quite easy to detect outliers; however, box plots, scatter plots, and histograms can be used to detect them in larger datasets.

Figure 5.1 shows a box plot of this data. Notice the presence of the circle at the top of the box plot denoting the outlier with a value of 52.

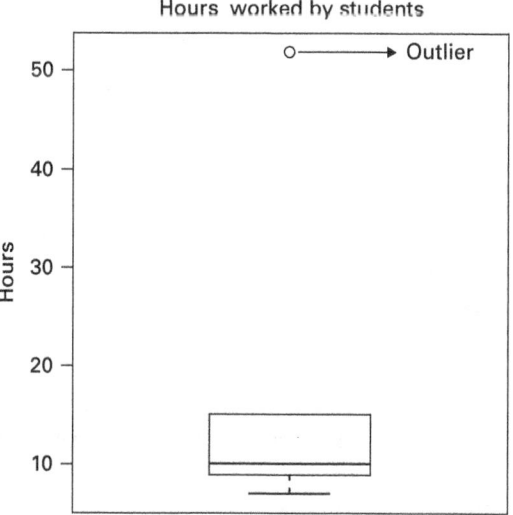

FIGURE 5.1. Box plot of student working data.

Alternatively, the scatter plot shown in Figure 5.2 shows data clustered at lower values with the exception of the single outlier at the top right of the figure.

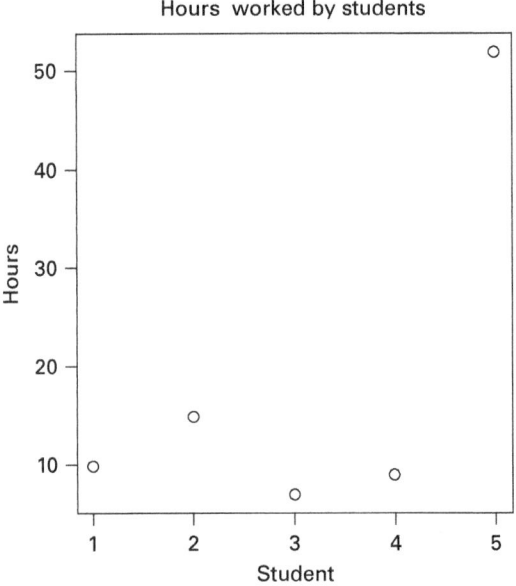

FIGURE 5.2. Scatter plot of student working data.

Finally, Figure 5.3 illustrates this same data with a histogram.

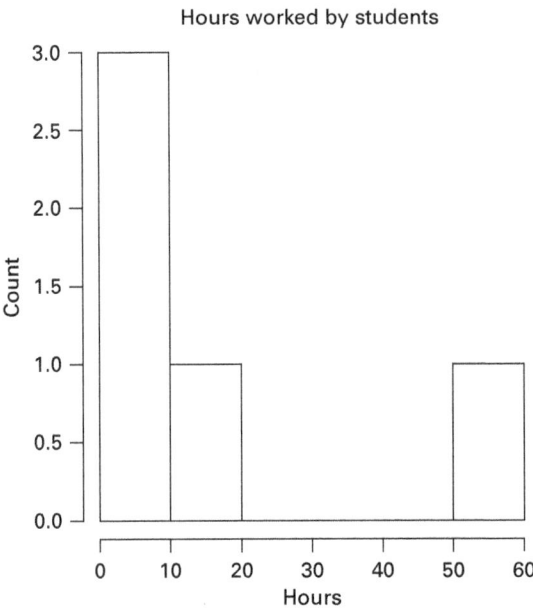

FIGURE 5.3. Histogram of student working data.

Even though the illustrated sample is small with only five observations, in all the graphs, we notice the presence of the outlier because of its physical distance from the remaining data.

In these examples, single variables were considered outliers; however, there are cases in which the relationships between variables can be outliers. Consider, for example, the scatter plot in Figure 5.4 that plots 10 students' overall GPAs with the average number of hours they study each week.

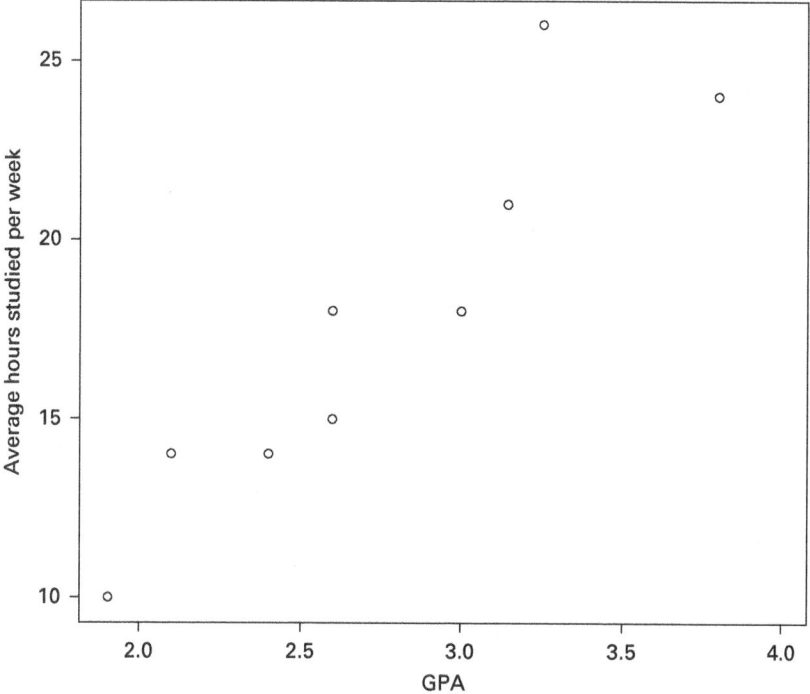

FIGURE 5.4. The relationship between study hours and GPA.

While the range of GPA scores goes from 1.9 to 4.0, the 4.0 is not an outlier because it is close to other observations (e.g., another student has a GPA of 3.8 and others are also above 3.0). Similarly, the number of hours studied is not an outlier, as another student also studies 10 hours and others study a similar number of hours too. What is unusual is the combination of such a high GPA with so few hours studied (how lucky is that person?!).

When looking for outliers, then, it is a good idea to plot the variables separately and together before making a decision about whether outliers exist in your data.

When outliers do exist, it might be useful to consider why this is the case. For instance, are there similar outliers in the population in a proportion similar to what appears in your sample? If this is the case, your outlying data may not be problematic. In general, outliers are more problematic when samples are small, the number of outliers is fairly large, or the outlying cases are more extreme.

- The sample is not "sufficiently large." As discussed previously, small samples are often underpowered, leaving inadequate opportunity to detect differences in samples when one actually exists in the population. Power analysis is often used in research to determine what sample size is needed to detect these differences. This is discussed in detail in the last chapter of this book.

## WHICH TEST OF TYPE I ERROR TO USE?

As stated earlier, the specific hypothesis test you will use is based on the level of measurement of your variables and how they are distributed. For the purposes of this book, we will limit our analyses to bivariate analyses, examining whether a relationship exists between two variables, as this forms the basis for more complex analyses. Table 5.2 can be used to determine which test to use given the level of measurement of each of your variables. Tests written in *italics* are nonparametric tests, whereas those written in regular type are parametric tests.

TABLE 5.2. Table of Statistical Tests

| Level of Measurement of First Variable | Level of Measurement of Second Variable | | |
|---|---|---|---|
| | Dichotomous | Nominal (3 or more categories) or ordinal | Interval or ratio |
| Dichotomous | Chi-square ($X^2$), Fisher's exact test, McNemar's test | Chi-square ($X^2$) | t-test, Mann-Whitney test, Wilcoxon signed rank test |
| Nominal (3 or more categories) or ordinal | | Chi-square ($X^2$) | One-way analysis of variance (ANOVA) followed by post hoc testing, Kruskal-Wallis test |
| Interval or ratio | | | Pearson's r, Spearman's rho (ρ) |

Consider the following research question: Is there a relationship between students' living arrangements and their majors? To answer this question, you would need to get information about where students lived and what they are studying. Perhaps living arrangements are operationalized categorically as living on campus in a dorm, living on campus in an apartment, or living off campus. Perhaps major is categorically operationalized as follows: natural sciences, behavioral/social sciences, education, engineering, mathematics, languages, art, or other liberal arts.

In this example, we have two variables, one that could be called *living* and the other we could call *major*. The *living* variable consists of three unranked categories, and the *major* variable consists of eight unranked categories. Using Table 5.2, you could determine that it would be appropriate to do a chi-square test to determine what relationship, if any, exists between those two variables.

Now think back to the experiment discussed earlier in this chapter where subjects were randomized to either Intervention A or TAU. If your research question was "Do those receiving Intervention A have different levels of self-esteem than those receiving TAU?" we would need to look at two variables as well. The *intervention* variable would be operationalized by subjects being assigned one of two values: Intervention A or TAU, which would be a dichotomous, nominal level of measurement. The *esteem* variable would be operationalized by recording each subject's actual self-esteem score on a standardized survey instrument. This would be an interval-level measure. Using Table 5.2, it would make sense to use either a *t*-test or one of the two listed nonparametric tests listed in that cell, depending on whether a parametric or nonparametric test is appropriate.

As you move through the remainder of this book, you will want to think about levels of measurement when you choose a statistical test, and, in addition, you will need to consider whether you can use a parametric test or, if you are unable to assume normality, a nonparametric test.

## CONCLUSION

In this chapter, you read about some of the foundational principles around hypothesis testing and statistical analysis as we moved from describing data to making inferences about a larger population from a sample. You learned that, when samples are sufficiently large, observations are independent of one another, and your sample does not have a lot of outliers, you can use parametric statistical tests. When this is not the case, nonparametric statistical tests are more appropriate.

The underlying principle behind statistical testing is that when observations in a sample are unusual, those observations may or may not be reflected in a larger population. The chances that they are not present in the larger population is a Type I error, and the chances of making that type of error are reflected in a *p*-value that can be calculated.

In quantitative analysis there is always a degree of uncertainty in our findings. Even if a small *p*-value is obtained, there is always a chance of the occurrence of a Type I error. In the behavioral sciences, a *p*-value of 0.05 is accepted to signify statistical significance and is an arbitrary minimum. Reporting the exact *p*-value is more helpful than merely stating that the result is significant at or below the 5%

level. While hypothesis testing can be a powerful tool in statistical analysis, it does have its limits. Statistical significance does not determine clinical or practical significance. Larger samples are more likely to produce statistical significance even when the effect is small. As a result, additional statistical analysis is needed, including effect sizes and confidence intervals, which can help inform decision making. Another method used to reduce uncertainty is power analysis, discussed in Chapter 10, which deals with the probability of not detecting a statistically significant difference when one exists. Another way of stating this is that statistical power is the probability of rejecting the null hypothesis when the null hypothesis is false.

## EXERCISES

1. For each of the following research questions, write a null hypothesis and either a directional or nondirectional alternate hypothesis.
   a. Does what dorm a student lives in have an impact on his or her grades?
   b. Does the number of roommates a student has impact his or her overall stress level?
   c. Do students with higher GPAs perceive themselves as better students?
   d. Are first-generation students similar to others in their current stress level?
   e. Do students in different colleges take similar course loads?
2. For each of the research questions in the previous set, find appropriate variables to operationalize each concept in the *student* dataset. Are these variables categorical or continuous? Name a test of Type I error from Table 5.2 that could be used to test the hypotheses you formed.
3. Why might it be difficult to use parametric hypothesis tests with the *student* dataset?
4. State whether each of the following is an example of a Type I error or a Type II error:
   a. Your doctor calls you. The rapid strep test that was negative yesterday was wrong! You actually have strep throat.
   b. It turns out all those brain training games you play online don't actually train your brain after all, but they're still fun.
   c. Even after a 99% compatibility score on Match.com, your date with Mr. Right turns out to be just plain wrong.
   d. Those prep classes that you took to improve your ACT score did nothing but take your parents' money!
   e. It turns out the DARE program you completed in fifth grade actually increased the number of kids taking drugs!

## *R* EXERCISE

1. Using the *hospital* dataset, determine whether there are outliers in any of the continuous variables. State what method you used to determine whether there were outliers.

# /// 6 ///    COMPARING SAMPLES

To work through the examples in this chapter, you will need to *install* and *load* the following *R* packages:

- *agricolae*
- *psych*
- *car*
- *effsize*

You should also load the following dataset into the Environment pane:

- *hospital*

For more information on how to do this, refer to Appendix C.

## INTRODUCTION

In this chapter, we will focus on hypotheses where we want to compare sample means to either a known population mean or other sample means. When we do this, we can test hypotheses regarding known groups of people with a continuous variable. However, to do this, we will have to consider another distribution, the *t*-distribution, which is preferable to the normal distribution for applying to these cases for several reasons. Therefore, we will begin this chapter by discussing why we want to use the *t* instead of the *z* most of the time. Then we will talk about how to apply the *t*-distribution to testing hypotheses related to comparing sample means to a population parameter or another sample.

## YET ANOTHER DISTRIBUTION

In previous chapters, we discussed how we could apply a known sampling distribution, such as the binomial or normal, to draw conclusions about whether observations we made were typical. When we wrote about these, we noted that there are conditions under which each could be applied. One of the conditions for using the normal distribution was that the sample was "sufficiently large" and it also required knowledge of the population standard deviation. However, what is "sufficiently large" is debatable and can differ under a variety of circumstances, and we often do not know the population standard deviation.

It is more practical, then, if we can make use of a sampling distribution that does not have those limitations. Luckily, the *t*-distribution exists! Like the normal distribution, the *t* is a class of distributions whose shape is similar to the normal distribution, but the exact shape is based on degrees of freedom. **Degrees of freedom**, or *df*, are the number of independent values that are assigned to a given distribution. They can be thought of as how much data can be missing while still being able to calculate a given value. Degrees of freedom are used to choose the most appropriate version of a probability distribution, and how degrees of freedom are calculated varies from test statistic to test statistic. *df* in determining the correct *t* statistic is defined as $n - 1$. Figure 6.1 displays the normal distribution and *t*-distributions based on four different degrees of freedom.

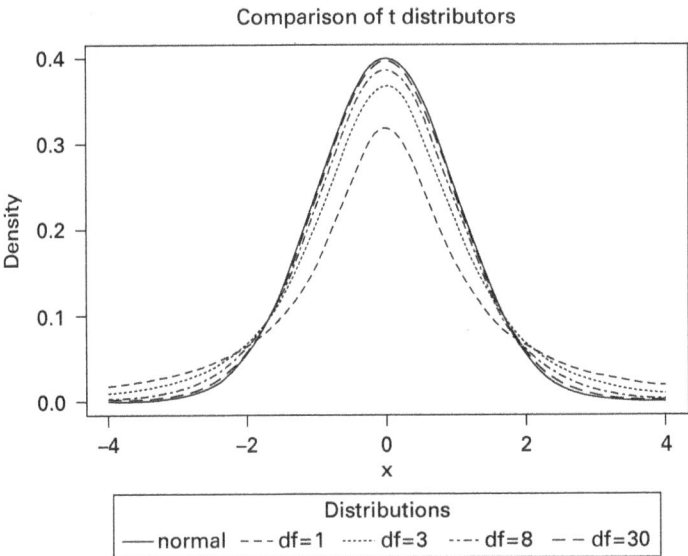

FIGURE 6.1. Normal distribution and *t*-distributions with four different degrees of freedom compared to the normal distribution.

Notice that, in general, with fewer degrees of freedom, the peak of the *t* is not as high as the normal, and the tails are fatter. However, as *df* approaches 29 (with a corresponding sample size of 30), the *t*-distribution mimics the shape of the normal distribution. This will provide a lot of flexibility as we proceed through this chapter because with the *t*, we are not necessarily limited by sample size, and we will therefore prefer using the *t* over the normal.

## ONE-SAMPLE *t*-TEST

Chapter 4 discussed the *z*-test, which compares a single sample mean with a known population mean to obtain the probability that the sample mean came from the population. The one-sample *t*-test achieves the same purpose. However, unlike the *z*-test, which uses the normal distribution, the one-sample *t*-test can be conducted on sample sizes less than 30, and the standard deviation of the population does not need to be known.

The formula for the *t*-test statistic uses the *t*-distribution discussed earlier and is shown as Formula 6.1:

$$\text{Formula 6.1:} \quad t = \frac{\bar{x} - \mu}{\frac{s}{\sqrt{n}}}$$

The denominator in this formula is called the **standard error of the mean** and is often notated as $s_{\bar{x}}$. The standard error of the mean helps by incorporating the sample size into the calculation of *t*. All other things being equal, larger sample sizes will result in lower standard errors, which will ultimately result in larger *t* values, since this is the denominator of the calculation of *t*. Therefore, larger sample sizes are more likely to result in "statistically significant" findings (i.e., *p*-values less than the criterion of significance).

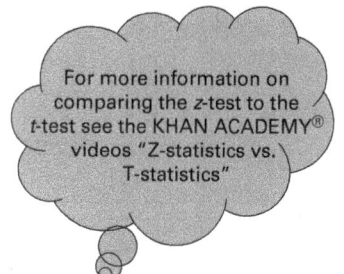

For more information on comparing the *z*-test to the *t*-test see the KHAN ACADEMY® videos "Z-statistics vs. T-statistics"

Let's use the example first shown in Chapter 2 of 10 students' use of social media to illustrate the use of the *t*-test statistic and distribution. In that chapter, we determined that the mean social media use for the sample was 3.8 hours with a standard deviation of 2.97 hours. We can compare the mean social media use of that sample to that of students across the country, who, on average, use social media for 4.3 hours per day. The null hypothesis in this case is that our students' social media use of 3.8 hours is not significantly different from that of other students at 4.3 hours. If the difference between the means is zero, we know the sample mean and population mean are equal; however, in this case, there is a difference of 0.5 hours.

A one-sample *t*-test can be applied to determine if we can reject the null hypothesis and therefore accept the alternative that the students at our university use social media differently than all other college students. A *z*-test would not be appropriate here because of the small sample size of 10, and we do not know the population standard deviation.

As is typical, we will set $\propto = 0.05$, and we need to calculate *df*, which for a one-sample *t*-test is $n - 1$. In this example, $df = n - 1 = 10 - 1 = 9$. Now, using Table 2 in Appendix F, we can come up with our critical value of 2.262 since we are using a nondirectional (i.e., two-tailed) hypothesis test. The hypothesis is nondirectional because we are only concerned whether students at our university use social media differently than students across the county; we are not concerned whether they specifically use it more or less. Figure 6.2 shows the *t*-distribution with 9 degrees of freedom and the areas beyond our critical value shaded. Calculated *t*-values in this shaded area would be considered unusual enough that our chances of making a Type I error would be less than 5%. If our calculated *t*-value was in one of these shaded areas, we would reject the null hypothesis and accept the alternate.

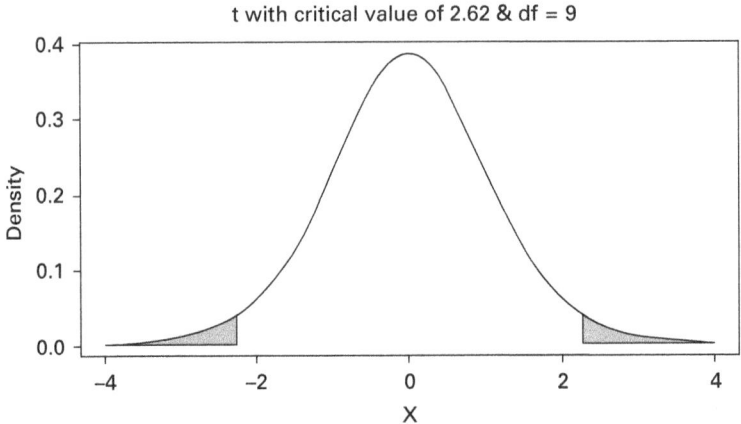

FIGURE 6.2. *t*-Distribution with a critical value of 2.262.

Now we can calculate the test statistic:

$$t = \frac{\bar{x} - \mu}{\frac{s}{\sqrt{n}}} = \frac{\bar{x} - \mu}{s_{\bar{x}}} = \frac{3.8 - 4.3}{\frac{2.97}{\sqrt{10}}} = \frac{-0.5}{\frac{2.97}{\sqrt{10}}} = \frac{-0.5}{\frac{2.97}{\sqrt{10}}} = \frac{-0.5}{\frac{2.97}{3.16}} = \frac{-0.5}{.94} = -0.532$$

Since our calculated *t* is NOT in the shaded area, we are unable to reject the null hypothesis. With our small sample of 10, it appears as if our students are not significantly different from other college students even though we observed lower levels of social media usage.

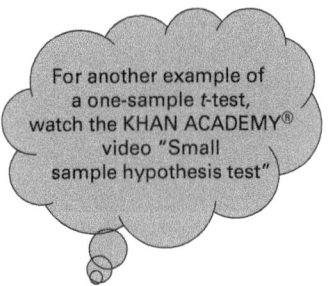

For another example of a one-sample *t*-test, watch the KHAN ACADEMY® video "Small sample hypothesis test"

## ONE TAILED OR TWO?

To expand on a topic previously discussed in Chapter 5, a nondirectional hypothesis is one that does not specify the nature of the relationship be-tween variables. When it comes to comparing means, a nondirectional hypothesis is one that only detects a difference between the observed (sample mean) and expected (population mean). In the previous example, this allows for our students to have significantly lower or higher usage than all other students.

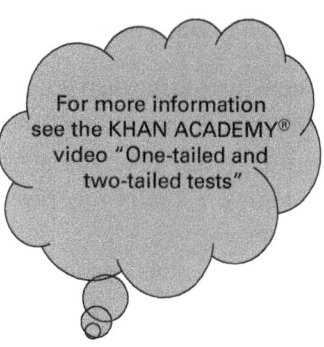

For more information see the KHAN ACADEMY® video "One-tailed and two-tailed tests"

On the other hand, a directional hypothesis specifies whether we believe our sample has a higher or lower mean than the population. For example, we could have hypothesized that our students use social media more than all other students. If the difference in the means is actually the opposite (i.e., all other students use social media more, which was the case), the null hypothesis could not be rejected.

## CALCULATING THE ONE-SAMPLE *T* IN *R*

Let us consider patients' length of stay (LOS) in the *hospital* dataset. We may want to know whether patients at County General have different lengths of stay from that of the national average of 6.3 days. In this case, the null hypothesis is that the true population mean for County General patients is 6.3, and the alternate, nondirectional hypothesis is that the true population mean is not 6.3. Enter the following into the Console to obtain the output displayed in Figure 6.3:

```
> t.test(hospital$LOS, mu=6.3)

             One Sample t-test

data:  hospital$LOS
t = 1.7097, df = 49, p-value = 0.09364
alternative hypothesis: true mean is not equal to 6.3
95 percent confidence interval:
  5.924688 10.955312
sample estimates:
mean of x
     8.44
```

FIGURE 6.3. Output from one-sample *t*-test.

Notice that the sample mean is entered first followed by the option **mu**, which is the population mean. The output shows the calculated test statistic, the degrees of freedom, and the probability of making a Type I error. In this case, we would not be able to reject the null hypothesis at the 0.05 level since the probability of making a Type I error is over 9%. The next line of the output states what the actual alternate hypothesis is, and the last two lines show what the sample mean is. Therefore, we know that the mean length of stay for the patients in our sample is 8.44 days.

## ANOTHER USE FOR THE $t$

There are times when we might want to compare a sample with a population; however, we do not know either the population mean or standard deviation. In these cases, we are generally not interested in testing a hypothesis, but we are interested in knowing a relatively narrow range within which the population parameter likely resides.

Consider this example: You wonder if children living in foster care exhibit more acting-out behavior than children who do not. To figure this out, you administer the Youth Behavior Scale to a sample of 16 children in foster care in your city. While you do not know the mean or standard deviation for the population, you do know that the mean for your sample is 52.5 and the $sd = 2.795$. We could figure out, with a given level of certainty, an interval in which the true population mean likely exists (i.e., the population parameter). If our sample were representative of the population of foster care children, our sample mean should be in this interval. Therefore, we will want to determine what the confidence interval is. Formula 6.2 displays how we would calculate this:

$$\text{Formula 6.2}: \quad \mu = \bar{x} \pm t_{\propto, df}\left(\frac{s}{\sqrt{n}}\right)$$

Before we begin, we need to determine how large we want our confidence interval to be. The more confident we want to be, the wider the confidence interval will be. To illustrate this, we will calculate both 95% and 99% confidence intervals for the previous example.

We can begin by calculating the 95% confidence interval, which is commonly used in the behavioral and social sciences. We will need to obtain the critical $t$ value given our $\propto$ and $df$. In this problem, $df = n - 1 = 16 - 1 = 15$. Since we want to be 95% sure that the population parameter is in the confidence interval, we are willing to take a 5% chance that we will be wrong (i.e., 100% certain $-$ 95% confident = 5% risk). Now, we can use Table 2 in Appendix F to obtain the critical $t$ of 2.131.

The lower range of the interval will use the difference part of the equation:

$$\mu = \bar{x} - t_{\propto, df}\left(\frac{s}{\sqrt{n}}\right) = 52.5 - (2.131)\left(\frac{2.795}{\sqrt{16}}\right) = 52.5 - (2.131)(0.70)$$
$$= 52.5 - 1.49 = 51.01$$

The upper range of the interval will use the summation part of the equation:

$$\mu = \bar{x} - t_{\propto, df}\left(\frac{s}{\sqrt{n}}\right) = 52.5 + (2.131)\left(\frac{2.795}{\sqrt{16}}\right) = 52.5 + (2.131)(0.70)$$
$$= 52.5 + 1.49 = 53.99$$

We can be 95% certain that the population mean is between 51.01 and 53.99. If we want to increase our certainty to 99%, we will need to use the critical $t$ value associated with 0.01 since 100% certain − 99% confident = 1% risk. Therefore, the only thing that will change in this problem will be $t_{\infty,df}$, which will increase from 2.131 to 2.947.

The lower range of the 99% confidence interval (CI) will use the difference part of the equation:

$$\mu = \bar{x} - t_{\infty,df}\left(\frac{s}{\sqrt{n}}\right) = 52.5 - (2.947)\left(\frac{2.795}{\sqrt{16}}\right) = 52.5 - (2.947)(0.70)$$
$$= 52.5 - 2.06 = 50.44$$

The upper range of the interval will use the summation part of the equation:

$$\mu = \bar{x} - t_{\infty,df}\left(\frac{s}{\sqrt{n}}\right) = 52.5 + (2.947)\left(\frac{2.795}{\sqrt{16}}\right) = 52.5 + (2.947)(0.70)$$
$$= 52.5 + 2.06 = 54.56$$

Notice that the 99% CI is wider, ranging from 50.44 to 54.56, than the 95% CI, which ranges from 51.01 to 53.99.

Because confidence intervals do not relay on a single value, these statistics are well suited to provide a more comprehensive approach to practical decision making.

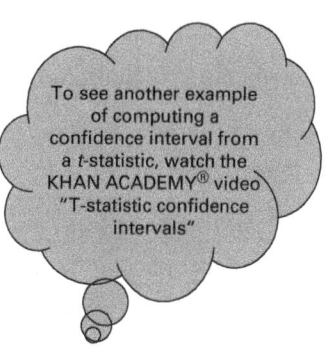

To see another example of computing a confidence interval from a $t$-statistic, watch the KHAN ACADEMY® video "T-statistic confidence intervals"

## COMPARING TWO-SAMPLE MEANS WITH AN INDEPENDENT SAMPLES $t$-TEST

The single-sample $t$-test can be extended to compare two samples. Very often, we are not interested in comparing a single sample to a population parameter, but we are interested in comparing the means between two different groups or samples. A very important assumption in this case is that the groups or samples must be independent of one another and that the samples have been selected randomly. Another important criterion is that the dependent variable is normally distributed. This rule can be relaxed if the sample size is 50 or greater. However, if these conditions are not met, nonparametric statistical tests will be necessary, and these are discussed in the next chapter.

As an example of how the independent samples $t$-test can be applied, consider two supplements that are designed to help people lose weight. In practice, the sample size in this example would be considered too small, but for the purposes of manual calculation, this example should suffice. An experiment is done in which five people are randomly assigned to use LOSEMORE for 30 days and another five are randomly assigned to use WEIGHLESS for 30 days. Our hypotheses are as follows:

$H_0: \mu_L = \mu_W$ and
$H_1: \mu_L \neq \mu_W$

After 30 days, we determine that $\bar{x}_L = 12$ and $\bar{x}_W = 10$. We also know that $s_L = 1.58$ and $s_W = 3.46$. The $t$-test for independent samples can be used to test this hypothesis; however, before we begin, we need to consider how similar the variances of each sample are. If they are similar, then we can pool the variances in our calculation of $t$. If they are different, then we will have to consider these separately.

So, before we proceed, we have to ask, "How can we determine if the variances are similar?" We can assume equal variances if the ratio of the larger to the smaller is less than 3.84, or, as illustrated in Formula 6.3:

$$\text{Formula 6.3}: \quad s_1^2 \sim s_2^2 \quad \text{if } F = \frac{s_{max}^2}{s_{min}^2} < 3.84$$

Let's begin by determining whether the variances of our samples are similar:

$$F = \frac{3.46^2}{1.58^2} = \frac{11.97}{2.5} = 4.79$$

Since the ratio of the variances is greater than 3.84, we will not be able to make use of the pooled variances. The formula for the $t$-test in this situation is shown in Formula 6.4:

$$\text{Formula 6.4}: \quad t = \frac{(\bar{X}_1 - \bar{X}_2)}{\sqrt{\frac{s_1^2}{N_1} + \frac{s_2^2}{N_2}}}$$

The denominator in this formula considers each variance separately. This is a more conservative calculation of the standard error of the difference, but it is the default of how $R$ calculates the independent samples $t$-test, as you will see later.

### Calculating Degrees of Freedom With Two Samples of Unequal Variances: Welch's Correction

A modification of the degrees of freedom is necessary when it cannot be assumed that the two populations being compared have the same standard deviations. Welch

(1938) suggested the following correction, which is a more conservative approach to calculating *df* when equal variances cannot be assumed. This calculation is shown in Formula 6.5.

In doing these calculations, it does not matter which sample you assign to be the first and which you assign to be the second:

$$A = \frac{s_1^2}{n_1} \quad \text{and} \quad B = \frac{s_2^2}{n_2}$$

$$\text{Formula 6.5:} \quad df = \frac{(A+B)^2}{\dfrac{A^2}{n_1 - 1} + \dfrac{B^2}{n_2 - 1}}$$

In continuing with our example:

$$A = \frac{s_1^2}{n_1} = \frac{2.5}{5} = 0.5$$

$$B = \frac{s_2^2}{n_2} = \frac{11.97}{5} = 2.39$$

$$df = \frac{(A+B)^2}{\dfrac{A^2}{n_1 - 1} + \dfrac{B^2}{n_2 - 1}} = \frac{(0.5 + 2.39)^2}{\dfrac{0.5^2}{5 - 1} + \dfrac{2.39^2}{5 - 1}} = \frac{8.35}{\dfrac{0.25}{4} + \dfrac{5.71}{4}} = \frac{8.35}{\dfrac{5.96}{4}} = \frac{8.35}{1.49} = 5.60 \cong 6$$

If we set $\propto$ at 0.05, using the calculated *df*, we can refer to Table 2 in Appendix F, and a *t* value of 2.571 or greater is necessary to reject the null hypothesis. That is, our calculated *t* should be in the shaded region of Figure 6.4 in order for us to reject the null and accept the alternate:

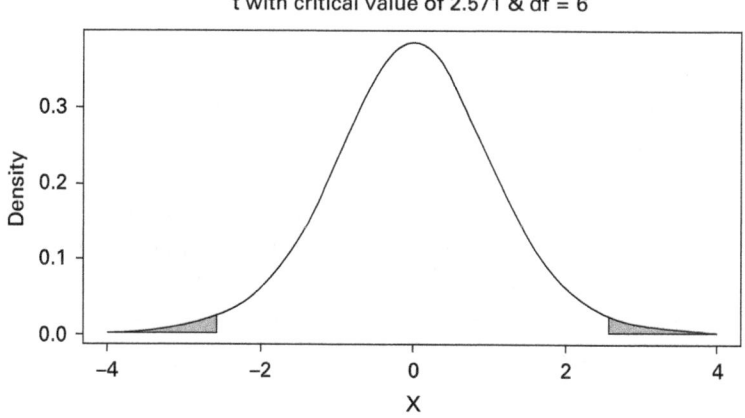

t with critical value of 2.571 & df = 6

FIGURE 6.4. *t* with critical value of 2.571 and *df* = 6.

$$t = \frac{(\bar{X}_1 - \bar{X}_2)}{\sqrt{\dfrac{s_1^2}{N_1} + \dfrac{s_2^2}{N_2}}} = \frac{(12-10)}{\sqrt{\dfrac{2.5}{5} + \dfrac{11.97}{5}}} = \frac{(12-10)}{\sqrt{0.5+2.39}} = \frac{2}{\sqrt{2.89}} = \frac{2}{1.7} = 1.18$$

Since our calculated $t$ is less than our critical $t$, we are unable to reject the null. Despite our observation that those who took LOSEMORE lost two more pounds than those who used WEIGHLESS, we cannot conclude that one is more effective than the other.

We could, however, try to determine with 95% confidence what the true difference in the means is between these weight loss remedies. This calculation is adapted from one we examined previously, Formula 6.2, and is shown in Formula 6.6:

$$\text{Formula 6.6:} \quad \mu_1 - \mu_2 = (\bar{X}_1 - \bar{X}_2) \pm t_{\infty, df}\sqrt{\frac{s_1^2}{N_1} + \frac{s_2^2}{N_2}}$$

We actually obtained all this required information in our $t$-test:

$$(\bar{X}_1 - \bar{X}_2) = 2 \text{ and}$$
$$t_{\infty, df} = 2.571 \text{ and}$$
$$\sqrt{\frac{s_1^2}{N_1} + \frac{s_2^2}{N_2}} = 1.7$$

Therefore, the lower bound of the 95% confidence interval is:

$$\mu_1 - \mu_2 = (\bar{X}_1 - \bar{X}_2) - t_{\infty, df}\sqrt{\frac{s_1^2}{N_1} + \frac{s_2^2}{N_2}} = 2 - 2.571(1.7) = 2 - 4.37 = 2.37$$

The upper bound of the 95% confidence interval is:

$$\mu_1 - \mu_2 = (\bar{X}_1 - \bar{X}_2) - t_{\infty, df}\sqrt{\frac{s_1^2}{N_1} + \frac{s_2^2}{N_2}} = 2 + 2.571(1.7) = 2 + 4.37 = 6.37$$

While we observed a two-pound difference in the means of these two weight loss treatments, we can be 95% certain that the true difference in the means ranges from –2.37 pounds (in which WEIGHLESS would be more effective than LOSEMORE) and 6.37 pounds (in which our observation that LOSEMORE is superior to WEIGHLESS would be true).

### The Independent *t*-Test When Variances Are Equal

Consider an example of an occupational therapist who wants to determine if patients receiving two different therapies, on average, score higher on an activity of daily living (ADL) scale administered at the end of their treatments. Finding out which therapy results in higher ADL scores will help the therapist choose the best therapy for his or her patients.

Tables 6.1 and 6.2 display the data and calculations for sample means and standard deviations.

TABLE 6.1. Data and Calculation of *s* for Patients Receiving Therapy A

| Received Therapy A | X (ADL) | $X - \bar{X}$ | $(X - \bar{X})^2$ |
|---|---|---|---|
| 1 | 8 | −0.3 | 0.09 |
| 2 | 7 | −1.3 | 1.69 |
| 3 | 9 | 0.7 | 0.49 |
| 4 | 8 | −0.3 | 0.09 |
| 5 | 9 | .7 | 0.49 |
| 6 | 7 | −1.3 | 1.69 |
| 7 | 10 | 1.7 | 2.89 |
| 8 | 9 | 0.7 | 0.49 |
| 9 | 9 | 0.7 | 0.49 |
| 10 | 7 | −1.3 | 1.69 |
| $\bar{X} = 8.3$ | $\Sigma X = 83$ | $\Sigma X - \bar{X} = 0$ | $(X - \bar{X})^2 = 13.19$ |

$$s = \sqrt{\frac{10.1}{9}} = \sqrt{1.12} = 1.06$$

TABLE 6.2. Data and Calculation of *s* for Patients Receiving Therapy B

| Received Therapy B | X (ADL) | $X - \bar{X}$ | $X - \bar{X})^2$ |
|---|---|---|---|
| 1 | 4 | −0.4 | 0.16 |
| 2 | 7 | 2.6 | 6.76 |
| 3 | 3 | −1.4 | 1.96 |
| 4 | 4 | −0.4 | 0.16 |
| 5 | 3 | −1.4 | 1.96 |
| 6 | 5 | 0.6 | 0.36 |
| 7 | 3 | −1.4 | 1.96 |
| 8 | 4 | −0.4 | 0.16 |
| 9 | 6 | 1.6 | 2.56 |
| 10 | 5 | 0.6 | .36 |
| $\bar{X} = 4.4$ | $\Sigma X = 44$ | $\Sigma X - \bar{X} = 0$ | $(X - \bar{X}^2) = 16.4$ |

$$s = \sqrt{\frac{16.4}{9}} = \sqrt{1.82} = 1.35$$

The question here is, "Is there a significant difference in ADLs between patients who receive Therapy A compared to Therapy B?" The ADL scale ranges from 1 to 10 with 1 representing the lowest score (i.e., those who are the least independent) and 10 the highest score (i.e., those who are the most independent).

In this case, our hypotheses are as follows:

$$H_0: \mu_A = \mu_B \text{ and}$$
$$H_1: \mu_A \neq \mu_B$$

The *t*-test for independent samples will be used to test this question; however, before we begin, we need to consider how similar the variances of each sample are. If they are similar, then we can pool the variances in our calculation of *t*. If they are different, then we will have to consider these separately:

$$F = \frac{1.35^2}{1.06^2} = \frac{1.82}{1.12} = 1.63$$

Since $F < 3.84$, we can assume equal variances and can make use of the pooled variance, and $df = n_1 + n_2 - 2 = 10 + 10 - 2 = 18$.

If we set $\propto$ at 0.05, using the calculated *df*, we can refer to Table 2 in Appendix F, and a *t* value of 2.101 or greater is necessary to reject the null hypothesis.

Now we will use the calculation for *t*, displayed in Formula 6.7, which makes use of pooling the variances of the samples:

$$\text{Formula 6.7}: \quad t = \frac{(\overline{X}_1 - \overline{X}_2)}{s_{\overline{x}_1 - \overline{x}_2}}$$

The denominator of this problem is referred to as the standard error of the difference, and the formula for this is shown in Formula 6.8:

$$\text{Formula 6.8}: \quad s_{(\overline{x}_1 - \overline{x}_2)} = s^2_{pooled} \sqrt{\left( \frac{1}{N_1} + \frac{1}{N_1} \right)}$$

where $s^2_{pooled}$ is defined in Formula 6.9:

$$\text{Formula 6.9}: \quad = s^2_{pooled} = \sqrt{\frac{(n_1 - 1)s_1^2 + (n_2 - 1)s_2^2}{n_1 + n_2 - 2}}$$

From the raw data provided in Tables 6.1 and 6.2, we have already calculated the means and standard deviations for each sample as follows:

$$\bar{X}_1 = 8.3 \quad \bar{X}_2 = 4.4$$

$$s_1 = 1.06 \quad s_2 = 1.35$$

$$(\bar{X}_1 - \bar{X}_2) = (8.3 - 4.4) = 3.9$$

and $n_1$ and $n_2$ are both 10.

Now we can calculate $s^2_{pooled}$ as follows:

$$s^2_{pooled} = \sqrt{\frac{(n_1-1)s_1^2 + (n_2-1)s_2^2}{n_1 + n_2 - 2}} = \sqrt{\frac{(10-1)1.06^2 + (10-1)1.35^2}{10+10-2}}$$

$$= \sqrt{\frac{(9)1.12 + (9)1.82}{18}} = \sqrt{\frac{10.08 + 16.38}{18}}$$

$$= \sqrt{\frac{26.46}{18}} = 1.21$$

And the standard error of the difference is:

$$s_{\bar{x}_1 - \bar{x}_2} = s^2_{pooled}\sqrt{\left(\frac{1}{N_1} + \frac{1}{N_1}\right)} = 1.21\sqrt{\left(\frac{1}{10} + \frac{1}{10}\right)} = 1.21\sqrt{0.2} = 0.54$$

And $t$ can (finally) be calculated:

$$t = \frac{(\bar{X}_1 - \bar{X}_2)}{s_{\bar{x}_1 - \bar{x}_2}} = \frac{3.9}{0.54} = 7.22$$

Since our calculated $t$ is greater than our critical value of 2.101, we are able to reject the null and accept the alternate. Treatment A is more effective than Treatment B in increasing ADLs, but we can enhance our understanding of this by determining the true difference in the population means by constructing a confidence interval.

Like calculations you have seen for confidence intervals in the past, Formula 6.10 makes use of the sample means, the critical test statistic, and the denominator of the calculated test statistic, in this case the standard error of the difference:

Formula 6.10: $\mu_1 - \mu_2 = (\bar{X}_1 - \bar{X}_2) \pm t_{\infty, df} s_{\bar{x}_1 - \bar{x}_2}$

Because we have already calculated the various elements of this formula, we can now calculate the true mean differences for Treatment A and Treatment B with 95% confidence:

$$\mu_1 - \mu_2 = (\bar{X}_1 - \bar{X}_2) \pm t_{\infty, df} s_{\bar{x}_1 - \bar{x}_2} = 3.9 \pm 2.101(0.54) = 3.9 \pm 1.13$$

Therefore:

$$2.77 \le \mu_1 - \mu_2 \le 5.03$$

Note that in this case, Treatment A is superior to Treatment B even at the low end of the difference.

## USING *R* TO CALCULATE THE INDEPENDENT SAMPLES *t*-TEST AND CONFIDENCE INTERVALS

The example data used in this section can be found in the *hospital* dataset. If you want to turn off scientific notation, enter the following in the Console:

```
>options(scipen=999)
```

Here, we will start building a profile of patients' LOS, which is simply a measure of how many whole nights patients spend in the hospital. We may, for instance, consider whether males or females stay in the hospital longer, on average. In this case:

$H_0 : \mu_M = \mu_F$ and
$H_A : \mu_M \ne \mu_F$

As discussed in the previous section, you will want to first determine whether the variances of length of stay are similar for our sample of males and our sample of females. To do this, enter the following in the Console to view the output in Figure 6.5:

```
> var.test(hospital$LOS~hospital$gender)
```

```
        F test to compare two variances

data:  hospital$LOS by hospital$gender
F = 0.23282, num df = 22, denom df = 26, p-value = 0.0009261
alternative hypothesis: true ratio of variances is not equal to 1
95 percent confidence interval:
 0.1037375 0.5375302
sample estimates:
ratio of variances
         0.2328153
```

FIGURE 6.5. `var.test()` function for *LOS* and *gender*.

In entering parameters into the `var.test()` function, notice that you enter the continuous variable first and separate it from the grouping variable that identifies the two samples with ~.

The output shows that the $p$-value for this test is less than 0.05, indicating that the variances for these samples are significantly different. This may seem a bit confusing since, in this case, the acceptance of the null hypothesis is desired. Nevertheless, and because the variances are significantly different, when we run the $t$-test, we will need to remember that we will need to specify unequal variances. With this in mind, you can run the $t$-test by entering the following into the Console to view the output displayed in Figure 6.6:

```
> t.test(hospital$LOS~hospital$gender)


        Welch Two Sample t-test

data:  hospital$LOS by hospital$gender
t = -2.2394, df = 38.735, p-value = 0.03094
alternative hypothesis: true difference in means is not equal to 0
95 percent confidence interval:
 -9.8266741 -0.4986078
sample estimates:
mean in group Female   mean in group Male
          5.652174             10.814815
```

FIGURE 6.6. *t*-Test comparing *LOS* for males and females.

The default in entering the $t$-test into $R$ is unequal variances, so you simply needed to enter the continuous variable separated by the categorical variable with two levels with ~.

In Figure 6.6, you will see the calculated $t$-statistic, the *df* based on Welch's correction, and the exact chances of making a Type I error. Since $p = 0.031$, we can reject the null and accept the alternate that males and females have different lengths of stay. In fact, looking at the bottom of this output, we see that, on average, females have an average LOS of 5.65 days compared to males with an average LOS of 10.81 days. The alternate hypothesis is that the difference in means is not equal to zero, and the 95% confidence interval shows that the true difference in means is likely between -9.83 days and -0.50 days. The CI interval shows that both values are negative because average male LOS was subtracted from average female LOS. Within the entire range of the 95% CI, females have shorter lengths of stay than males, which is related to the differences in means being significantly different.

As another example, we may want to determine if patients returning to the hospital within 30 days (*readmit30*) had a more extended stay in the hospital (LOS) than those who were not readmitted within 30 days. In this case:

$H_0: \mu_R = \mu_{NR}$ and
$H_A: \mu_R \neq \mu_{NR}$

Like our earlier example, we need to first determine whether the variances between these samples are equal by entering the following into the Console:

```
> var.test(hospital$LOS~hospital$readmit30)
```

The results are displayed in Figure 6.7.

```
         F test to compare two variances

data: LOS by readmit30
F = 2.109, num df = 29, denom df = 19, p-value = 0.09332
alternative hypothesis: true ratio of variances is not equal to 1
95 percent confidence interval:
 0.8780477 4.7058015
sample estimates:
ratio of variances
          2.10902
```

FIGURE 6.7. Examining the variances in LOS between those readmitted and those not readmitted within 30 days of discharge.

Since the *p*-value shown is greater than 0.05, we assume that the variances are equal when we run our *t*-test by entering the following into the Console to get the output shown in Figure 6.8:

```
> t.test(hospital$LOS~hospital$readmit30, var.
equal=T)
```

```
         Two Sample t-test

data: LOS by readmit30
t = -0.49192, df = 48, p-value = 0.625
alternative hypothesis: true difference in means is not equal to 0
95 percent confidence interval:
 -6.443934  3.910600
sample estimates:
 mean in group No mean in group Yes
         7.933333          9.200000
```

FIGURE 6.8. *t*-Test examining relationship between LOS and readmission to the hospital within 30 days of discharge.

In entering the `t.test()` function, notice that we have added the option `var.equal=T` to indicate that the variances are not significantly different from one another. In the output, you will see that the average length of stay for those who are not ultimately readmitted is 7.93 days compared to 9.2 days for those who are readmitted. These mean differences are not statistically different at the 0.05 level since our calculated *p*-value is greater than that at 0.625. Our 95% CI shows a true mean difference ranging from -6.44 days to 3.91 days. Even though we notice that those readmitted spend more time in the hospital than those not readmitted, we are unable to reject our null hypothesis and conclude that there is no difference in the population means of these groups.

## RELATED SAMPLES/DEPENDENT SAMPLES *t*-TEST

On occasion, it is necessary to evaluate the same subjects at two different time points such as before and after an intervention. In another type of study, subjects in a treatment group may be matched to a control group on a set of characteristics. Because these lack independence, a special form of the *t*-test must be utilized, which uses the difference in scores for each of the pairs.

The null hypothesis would, therefore, be that the mean difference between the pairs is zero. For example, a social worker in a residential care facility is interested in evaluating whether an intervention to improve acclimation of new residents is effective. Ten residents are assessed for acclimation before and after an intervention. The acclimation scale ranges from 1 (lowest level of acclimation) to 20 (highest level of acclimation). The data are presented in Table 6.3. In reality, 10 subjects would be too small a sample to do a *t*-test (i.e., you would prefer a nonparametric test), but for illustration purposes and ease of manual calculation, we will proceed.

TABLE 6.3. Data and Preliminary Calculation for Paired *t*-Test

| Subject | Acclimation Before Intervention | Acclimation After Intervention | After–Before $X_D$ | $\Sigma(X_D - \bar{X}_D)$ | $\Sigma(X_D - \bar{X}_D)^2$ |
|---|---|---|---|---|---|
| 1 | 12 | 15 | 3 | .9 | .81 |
| 2 | 9 | 9 | 0 | −2.1 | 4.41 |
| 3 | 16 | 18 | 2 | −0.1 | 0.01 |
| 4 | 12 | 19 | 7 | 4.9 | 24.01 |
| 5 | 15 | 17 | 2 | −0.1 | 0.01 |
| 6 | 17 | 19 | 2 | −0.1 | 0.01 |
| 7 | 11 | 15 | 4 | 1.9 | 3.61 |
| 8 | 14 | 11 | −3 | −5.1 | 26.01 |
| 9 | 18 | 20 | 2 | −0.1 | 0.01 |
| 10 | 15 | 17 | 2 | −0.1 | 0.01 |

$$\Sigma X_D = 21; \bar{X}_D = \Sigma \frac{X_D}{N} = \frac{21}{10} = 2.1 \qquad \Sigma = 58.9$$

The critical *t* will be based on *df*, which is $df = n - 1 = 10 - 1 = 9$. Setting $\alpha$ at 0.05, our threshold for rejecting the null will be $t \geq 2.262$.

The formulae for both the standard deviation of the paired difference and the dependent sample *t* are displayed in Formulae 6.11 and 6.12, respectively:

$$\text{Formula 6.11:} \quad s_D = \sqrt{\frac{\Sigma(X_D - \bar{X}_D)^2}{n-1}}$$

$$\text{Formula 6.12:} \quad t = \frac{\bar{X}_D}{\frac{s_D}{\sqrt{n}}}$$

The first step is to calculate the mean difference of the scores by subtracting the acclimation score after the intervention from the acclimation score before the intervention. Summing that difference and dividing by the total provides the mean difference, which becomes the numerator in the expression for the *t*-test. Next, $s_D$ is calculated as follows. Begin by entering the sum of the squared deviation from the mean difference shown in Table 6.3:

$$s_D = \sqrt{\frac{\Sigma(X_D - \bar{X}_D)^2}{n-1}} \, 2.334 = \sqrt{\frac{58.9}{9}} = \sqrt{6.54} = 2.56$$

Now calculate *t* as follows:

$$t = \frac{\bar{X}_D}{\frac{s_D}{\sqrt{n}}} = \frac{2.1}{\frac{2.56}{\sqrt{10}}} = \frac{2.1}{\frac{2.56}{3.162}} = \frac{2.1}{0.8096} = 2.59$$

The $df = n - 1 = 10 - 1 = 9$.

Based on our calculated test statistic, there was statistically significant improvement in acclimation after the introduction of the intervention since $2.59 > 2.262$.

## USING *R* TO CALCULATE THE DEPENDENT SAMPLES *t*-TEST

The example data used in this section can be found in the *hospital* dataset. Here we will look at how memory changes in hospital patients. The variable *membefore* assessed patient memory at admission, and *memafter* assessed memory just before discharge. Since these two variables are not independent (i.e., they are both measures of memory on the same individuals), a dependent samples *t*-test is appropriate to test the following hypotheses:

$H_0: \mu_B = \mu_A$ and
$H_1: \mu_B \neq \mu_A$

Before we conduct our *t*-test, however, we might want to better understand each variable. To get some summary statistics on each, enter the following into the Console:

```
>summary(hospital$membefore)
Min. 1st Qu.  Median   Mean 3rd Qu.    Max.
9.00   13.25   16.00   15.62   17.75   24.00

>summary(hospital$memafter)
  Min. 1st Qu.  Median   Mean 3rd Qu.    Max.
     8      15      18     17     20      23
```

Notice that the mean for patients' memory at admission is 15.62 and at discharge it was higher at 17.

Enter the following into the Console to run the *t*-test:

```
>t.test(hospital$membefore,hospital$memafter,
paired=TRUE)
```

Notice the inclusion of the option **paired=TRUE**, which signifies the use of the dependent samples *t*-test. The results are displayed in Figure 6.9.

```
        Paired t-test

data:  hospital$membefore and hospital$memafter
t = -2.1867, df = 49, p-value = 0.03357
alternative hypothesis: true difference in means is not equal to 0
95 percent confidence interval:
 -2.6482276 -0.1117724
sample estimates:
mean of the differences
              -1.38
```
FIGURE 6.9. *R* example of dependent sample *t*-test.

The difference in the means in this sample was –1.38, and there was a statistically significant improvement in memory scores by patient discharge. Notice that the confidence interval included in the output noted that we can be 95% confident that the true difference in means is somewhere between –2.65 and –0.111.

## HOW BIG IS THE DIFFERENCE?

In the examples presented previously, there was some observed change from the first time point to the second. The hypothesis test was able to tell us if there were significant mean differences between these two time points, and the confidence interval was able to give us a strong sense as to what the true difference in means was for the population. Very often, however, we may want to describe how much change there is, and, as discussed previously, hypothesis testing cannot help with this, as sample size greatly impacts the threshold for determining statistically significant change.

It is useful, however, to be able to say something like, "On average, people who engaged in Intervention X saw a 30% improvement in their grades." One way to quantify this is through a class of descriptive statistics called **effect sizes**. It is often helpful to quantify how large of a difference exists between two independent groups or at two time points for dependent samples, as this can suggest the clinical or practical significance of observed differences separate from statistical significance.

In this section, we will cover one of the most popular measures of effect size, Cohen's $d$. This measure of effect size is widely used because it accounts for differences in variation between groups. The formula is presented as Formula 6.13:

$$\text{Formula 6.13:} \quad d = \frac{(\bar{X}_1 - \bar{X}_2)}{s_{pooled}}$$

where

$$s_{pooled} = \sqrt{\frac{(n_1 - 1)s_1^2 + (n_2 - 1)s_2^2}{n_1 + n_2 - 2}}$$

As an example, let's apply this to the occupational therapy patients receiving Therapy A compared to Therapy B described in Tables 6.1 and 6.2. Based on our previous calculations, we know that:

$$(\bar{X}_1 - \bar{X}_2) = 8.3 - 4.4 = 3.9 \text{ and } s_{pooled} = \sqrt{1.472} = 1.213$$

Therefore:

$$d = \frac{(\bar{X}_1 - \bar{X}_2)}{s_{pooled}} = \frac{3.9}{1.213} = 3.125$$

The interpretation of Cohen's $d$ is based on $z$-scores. The score represents the degree of difference in ADLs for those receiving Treatment A compared to those receiving Treatment B. An effect size of 3.2 denotes a little over a three standard deviation difference between the groups. Therefore, an effect size of 0 shows no difference, while an effect size of 3.21 indicates that those who received Treatment A had, on average, 49.94% better ADLs than those who received Treatment B. This value was obtained from the $z$-score table; a $z$ of 3.22 corresponds to 49.94% of the area under the standard normal distribution from the mean to 3.22 standard deviations (0.50 to 0.0006).

In terms of interpreting Cohen's $d$ qualitatively, a small effect size is loosely defined as $d = 0.2$, a medium effect is $d = 0.5$, and a large effect is $d = 0.8$ (Cohen, 1988). Thinking about this statistically, this means that a 49.94% difference between the groups is large, and likely has practical implications. Patients receiving Treatment A probably have visibly better self-care skills than those receiving Treatment B.

Cohen's $d$ can also be applied to the paired $t$-test to compare the degree of change from one time period to the next. The formula would be the same as presented earlier where the numerator is the mean of the difference of the pairs. This is illustrated in Formula 6.14:

$$\text{Formula } 6.14: \quad d = \frac{\bar{X}_D}{s_D}$$

As an example, we can now consider how well new residents acclimated to their environment based on the data shown in Table 6.3. We know from our calculations that $\bar{X}_D = 2.1$ and $s_D = 2.56$. Therefore:

$$d = \frac{\bar{X}_D}{s_D} = \frac{2.1}{2.56} = 0.820$$

This means that we saw a large degree of change from the time residents moved in to when their acclimation was assessed again. This is great news! Whatever is being done to help acclimate new residents may be having a positive effect!

## USING *R* TO CALCULATE EFFECT SIZE

To compute and interpret Cohen's $d$ in *R*, you will need to install and require the *effsize* package to be available on CRAN. Once this is done, effect size for the memory difference in hospital patients can be calculated as displayed in Figure 6.10. Enter the following in the Console:

```
> cohen.d(hospital$membefore, hospital$memafter,
paired=T)
```

```
Cohen's d

d estimate: -0.3092439 (small)
95 percent confidence interval:
          inf         sup
-0.70850259  0.09001475
```

FIGURE 6.10. Effect size for memory changes.

Notice that we included the option `paired=T` since the samples were dependent and the data were paired. In this case, the calculated value for Cohen's *d* is -0.309. The confidence intervals indicate that it is 95% likely that the true effect size is between -0.709 and 0.090. As mentioned, the interpretation of Cohen's *d* is based on *z*-scores. The score then represents the degree of average improvement in the postintervention period over the preintervention period. The calculated effect size denotes less than a one standard deviation difference and is considered a small effect, as shown in the output. While we noted statistically significant improvement in memory in hospital patients, with a small effect size, those differences may be minimally evident.

For independent samples, the `cohen.d()` function would simply exclude the `paired=T` option as the default in *R* is that the samples are independent.

## ONE-WAY ANALYSIS OF VARIANCE

Throughout this chapter we have assessed samples by comparing mean differences between samples. These tests required a categorical variable with two samples or groups and a continuous (i.e., interval or ratio level) dependent variable. In this section we will discuss analysis of variance, which allows for the comparison of more than two samples.

As an example, the psychology department in your school is wondering if students entering their major from different regions of the country come in with the same base knowledge. To test this, they randomly recruit 30 students to take an entrance test on the basics of psychology. Table 6.4 displays the raw data from each of these sections.

TABLE 6.4. Test Grades for Students From Three Regions

| Student # | Region | Raw Score |
|---|---|---|
| 1 | East Coast | 14 |
| 2 | East Coast | 15 |
| 3 | East Coast | 12 |
| 4 | East Coast | 9 |
| 5 | East Coast | 10 |
| 6 | East Coast | 16 |
| 7 | East Coast | 9 |
| 8 | East Coast | 10 |
| 9 | East Coast | 11 |
| 10 | East Coast | 14 |

TABLE 6.4. Continued

| Student # | Region | Raw Score |
| --- | --- | --- |
| 11 | Midwest | 12 |
| 12 | Midwest | 12 |
| 13 | Midwest | 14 |
| 14 | Midwest | 9 |
| 15 | Midwest | 17 |
| 16 | Midwest | 14 |
| 17 | Midwest | 14 |
| 18 | Midwest | 17 |
| 19 | Midwest | 15 |
| 20 | Midwest | 12 |
| 21 | West Coast | 19 |
| 22 | West Coast | 17 |
| 23 | West Coast | 16 |
| 24 | West Coast | 14 |
| 25 | West Coast | 18 |
| 26 | West Coast | 16 |
| 27 | West Coast | 17 |
| 28 | West Coast | 18 |
| 29 | West Coast | 19 |
| 30 | West Coast | 17 |

One might think it logical to conduct multiple $t$-tests comparing those coming from the East Coast to those from the Midwest, those coming from the Midwest to those from the West Coast, and those coming from the East Coast to those from the West Coast.; however, conducting multiple $t$-tests increases the likelihood of making a Type I error. At the 0.05 level, after running 20 $t$-tests that show statistically significant differences, at least one will lead us to reject the null and accept the alternate incorrectly. Therefore, the chances of a Type I error after running three tests will be greater than if only one test is conducted.

With analysis of variance (ANOVA), a single test is conducted to compare more than two samples simultaneously and the $F$-statistic is used to test hypotheses. This procedure has a couple of important assumptions: The **residuals** (i.e., the difference between what is observed and what is predicted) are normally distributed in the population and the variance of the residuals is the same across the values of the independent variable (i.e., region of the country). In ANOVA, variances are compared, not means. Residuals are discussed in more detail in Chapter 8.

In our example, since all students in each sample were given the same exam, the assumption is that the variances within each region of the country are only due to random error and are homogeneous.

On the other hand, it is assumed that differences between student scores are due to the difference in region of the country from which students come. Therefore, the null hypothesis is as follows: The sample means are equal to the means in the population. To test this, ANOVA compares the amount of variance between groups (i.e., region to region) to the variance within groups (i.e., students from each region).

The calculation for the $F$-test is displayed in Formula 6.15. It is the ratio between the mean square between groups and the mean square within groups. Several steps need to occur before this final step is reached:

$$\text{Formula 6.15:} \quad F = \frac{MS_{bet}}{MS_{with}}$$

In our example, each group has its own mean, $\bar{X}_E, \bar{X}_M$, and $\bar{X}_W$, for exam score. A mean for all three groups can be calculated, which is known as the total mean or grand mean, denoted as $\bar{X}_{tot}$ or $\bar{X}$. We can compare each individual score to the total mean ($\bar{X}_{tot}$) of 14.23. Let's look at the fourth student in Table 6.4, whose score is 9. This student deviates from the grand mean as follows:

$$\bar{X}_{tot} - X_4 = 14.23 - 9 = -5.23$$

The deviance of this student's score from his or her own group's mean, which is known as the within-group deviation, can be calculated as follows:

$$\bar{X}_E - X_4 = 12 - 9 = 3$$

Finally, a *between-group* deviation can be calculated for East Coast students' mean by subtracting the grand mean from the group mean as follows:

$$\bar{X}_{tot} - \bar{X}_4 = 14.23 - 12 = 2.23$$

To begin our calculations, we will look at the data for each region separately. Table 6.5 displays summary statistics for each.

TABLE 6.5. Deviation Scores Comparing Exam Scores
Between Professors

| East Coast | | Midwest | | West Coast | |
|---|---|---|---|---|---|
| X | X² | X | X² | X | X² |
| 14 | 196 | 12 | 144 | 19 | 361 |
| 15 | 225 | 12 | 144 | 17 | 289 |
| 12 | 144 | 14 | 196 | 16 | 256 |
| 9 | 81 | 9 | 81 | 14 | 196 |
| 10 | 100 | 17 | 289 | 18 | 324 |
| 16 | 256 | 14 | 196 | 16 | 256 |
| 9 | 81 | 14 | 196 | 17 | 289 |
| 10 | 100 | 17 | 289 | 18 | 324 |
| 11 | 121 | 15 | 225 | 19 | 361 |
| 14 | 196 | 12 | 144 | 17 | 289 |
| $\Sigma\,120$ | $\Sigma\,1500$ | $\Sigma\,136$ | $\Sigma\,1904$ | $\Sigma\,171$ | $\Sigma\,2945$ |
| $\bar{X}_e = 12.0$ | | $\bar{X}_m = 13.6$ | | $\bar{X}_w = 17.1$ | |

$$\sum X_{tot} = 427$$

$$\bar{X}_{tot} = 14.23$$

ANOVA does not partition deviations but rather variances by calculating the sum of squares between groups and within groups. As a result, before the mean squares between and within groups can be calculated, the between-group sum of squares and the within-group sum of squares must be derived first. These sums of squares are the average of the squared deviation scores.

The formula and calculation of the between-group sum of squares are displayed in Formula 6.16:

$$\text{Formula 6.16:}\quad SS_{bet} = \sum \frac{(\sum X_g)^2}{n_g} - \frac{(\sum \bar{\bar{X}})^2}{N}$$

$$= \frac{(120)^2}{10} + \frac{(136)^2}{10} + \frac{(171)^2}{10} - \frac{(427)^2}{30}$$

$$= \frac{14400}{10} + \frac{18496}{10} + \frac{29241}{10} - \frac{182329}{30}$$

$$= 1440 + 1849.6 + 2924.1 - 6077.63$$

$$= 136.07$$

The calculation and formula for the sum of squares within groups are displayed in Formula 6.17:

$$\text{Formula 6.17:} \quad SS_{with} = \Sigma X^2 - \Sigma \frac{(EX_g)^2}{n_g}$$

$$= 6349 - \frac{14400}{10} + \frac{18496}{10} + \frac{29241}{10}$$

$$= 135.3$$

Once the sum of squares is computed, to obtain the between-group mean square, the between-group sum of squares is divided by the between-group's degrees of freedom defined as $df_{with} = N$ of subjects $-$ Number of groups $= N - k$. In our example, $df_{with} = 30 - 3 = 27$. To obtain the within-group mean square, the within-group sum of squares is divided by the within-group degrees of freedom defined as $df_{bet} = n$ of groups $- 1 = k - 1$. In our example, $df_{bet} = 2$. The calculation and example for the mean square between groups is shown in Formula 6.18:

$$\text{Formula 6.18:} \quad MS_{bet} = \frac{SS_{bet}}{df_{bet}} = \frac{136.07}{2} = 68.04$$

Similarly, the calculation and example for the mean square within groups is shown in Formula 6.19:

$$\text{Formula 6.19:} \quad MS_{with} = \frac{SS_{with}}{df_{with}} = \frac{135.3}{27} = 5.01$$

The final step is the calculation of the $F$-statistic, using Formula 6.15, which is shown as follows:

$$F = \frac{MS_{bet}}{MS_{with}} = \frac{68.04}{5.01} = 13.58$$

Refer to the table for the critical values for the $F$-statistic—Table 3 in Appendix F. We will use α set at 0.05. The between-group degrees of freedom, which correspond to the numerator in the calculation of $F$, are across the top row, and the within-group degrees of freedom, which correspond to the denominator in the calculation of $F$, are along the first column. With within-group degrees of freedom of 2 and 27, an $F$-value of 3.35 or greater is needed to reject the null hypothesis. The calculated $F$ of 13.58 exceeds the minimum value needed, so we can conclude that that region of the country is, indeed, related to exam scores.

This example had what is known in statistics as balanced groups, meaning that the number of observations in each sample is equal. Often this is not the case and a different form of ANOVA is utilized. Luckily for us, $R$ automatically adjusts for unequal group sizes.

## ONE-WAY ANALYSIS OF VARIANCE USING $R$

The example data used in this section can be found in the *hospital* dataset. In our sample of patients, each was primarily cared for by one of three health care professionals: an MD, an RN, or a PA (i.e., physician's assistant). Upon discharge, the patient was asked to complete a scale assessing how sensitive he or she thought the primary caregiver was to his or her overall needs. The scale ranged from 1 (not sensitive at all) to 20 (perfectly sensitive). Who cared for the patient is the variable labeled *position*, and the sensitivity score is the variable labeled *sensitivity*. We could hypothesize that there were differences in sensitivity based on who was caring for the patient, and the null hypothesis would be that there would be no difference in sensitivity based on position.

To see Sal Khan explain ANOVA, watch the KHAN ACADEMY® videos "ANOVA 1: Calculating SST (total sum of squares," "ANOVA 2: Calculating SSW and SSB (total sum of squares within and between)," and "ANOVA 3: Hypothesis test with F-statistic"

To begin, we may want to simply look at the means and standard deviations of each of these groups. Load and require the *psych* package to begin this analysis, and then enter the following into the Console:

```
> describeBy(hospital$sensitivity, hospital$position)
```

The output from this function is displayed in Figure 6.11.

```
Descriptive statistics by group
group: MD
   vars  n  mean   sd median trimmed  mad min max range skew kurtosis   se
X1    1 17 13.41 3.08     14   13.33 4.45   9  19    10 0.06    -1.37 0.75
--------------------------------------------------------------
group: PA
   vars  n  mean   sd median trimmed  mad min max range  skew kurtosis   se
X1    1 16 13.88 2.16     14      14 2.97   9  17     8 -0.37    -0.54 0.54
--------------------------------------------------------------
group: RN
   vars  n  mean   sd median trimmed  mad min max range  skew kurtosis   se
X1    1 17 16.06 2.79     17   16.33 1.48   9  19    10 -0.98     0.04 0.68
```

FIGURE 6.11. Looking at sensitivity across three types of caregivers.

It seems that, on average, nurses scored higher than both MDs and PAs in overall sensitivity, although there was slightly more variability in the RN group than in the PA group. From this data, could we consider that RNs in general are more sensitive than the others? Since there are three samples in this hypothesis, we will need to use ANOVA to test it. To begin, we will create a vector to contain the ANOVA, and we will use the **aov()** function:

```
> a1<-aov(hospital$sensitivity~hospital$position)
```

Notice that the continuous variable is listed first in the function, and it is separated by the categorical variable with ~.

To view the results, displayed as Figure 6.12 and saved as **a1**, simply enter the following into the Console:

```
> summary(a1)

                   Df Sum Sq Mean Sq F value Pr(>F)
hospital$position  2   67.6   33.81   4.581 0.0152 *
Residuals         47  346.8    7.38
---
Signif. codes:  0 '***' 0.001 '**' 0.01 '*' 0.05 '.' 0.1 ' ' 1
```

FIGURE 6.12. Example of one-way ANOVA output using *R*.

You will notice that the inputs for the "within" calculations are on the top row and for the "between" calculations are on the bottom row. The total calculated *F* is displayed, as are the chances of making a Type I error.

As we mentioned earlier, one of the assumptions of ANOVA is equality of variance. This assumption can be tested in *R* utilizing Levene's test. First, you will need to install and load the *car* package to unitize the levene.test() function. The syntax is displayed as follows:

```
>leveneTest(a1)
```

Notice that we just have to enter the vector *a1* between the parentheses and the following is displayed in the Console:

```
Levene's Test for Homogeneity of Variance
(center = median)
     Df F value Pr(>F)
group 2 1.2761 0.2886
47
```

Here the desired outcome is a $p$-value greater than 0.05, indicating that the variances between the groups are not significantly different. Accepting the null hypothesis in this case means that we can accept the hypothesis of equal variances between groups.

Looking at the ANOVA and Levene's test together, it seems that position does matter in overall sensitivity, but now we have another question: Where are those differences? Are nurses different from both PAs and MDs? Are MDs and PAs different from each other?

## WHERE IS THE DIFFERENCE? USING POST HOC TESTS

Unlike the *t*-test, when an *F*-test is significant we know there is a statistical difference between group means, but not where the differences arise. In the previous example, differences could occur between MDs and PAs, MDs and RNs, and/or PAs and RNs. One, two, or three pairs could be significantly different from one another. To know which pairs are significantly different, each pair must be tested individually. This is exactly the purpose of a post hoc test. A post hoc test pinpoints which group differences account for a significant *F*. It compares all group pairs to ascertain which are significantly different.

Obviously, it is not necessary to do a post hoc test if the *F*-test is not statistically significant. There are a number of these types of tests available. The two most popular are the Tukey honestly significant difference test (Tukey HSD) and the Scheffe, which will be covered in this section. However, because of the complexities involved in these calculations, we will only demonstrate how to do this in *R* with a continuation of the example we began earlier as an illustration.

### Tukey Honestly Significant Difference Test

Since we did observe a significant *F*-test, we can proceed with the Tukey HSD. Calculating the Tukey HSD is a simple one-line statement and is displayed as follows:

```
>TukeyHSD(a1)
```

```
         Tukey multiple comparisons of means
           95% family-wise confidence level

Fit: aov(formula = hospital$sensitivity ~ hospital$position)

$`hospital$position`
           diff        lwr      upr     p adj
PA-MD 0.4632353 -1.8266085 2.753079 0.8765917
RN-MD 2.6470588  0.3921766 4.901941 0.0178724
RN-PA 2.1838235 -0.1060202 4.473667 0.0644423
```

FIGURE 6.13. Using *R* to calculate the Tukey HSD.

Notice the ANOVA vector *a1* is included between the parentheses. The results displayed in Figure 6.13 confirm the hand calculations in the previous section. The column labeled *diff* displays the difference in means between the groups, while the columns labeled *lwr* and *upr* are the bounds of a 95% confidence interval for the mean differences in the groups. The *p adj* column displays the significance level for the difference in the pairs. At the 0.05 level, RNs are thought to be significantly more sensitive compared to MDs. On the other hand, PAs and MDs have similar levels, while the differences between RNs and PAs are approaching significance.

### The Scheffe

Another well-known post hoc test is the Scheffe, which is the most conservative of this class of tests. The test compares all pairs but also contrasts more than two means at a time. Because it is a more conservative test, it is more likely that a Type II error will occur (i.e., concluding there is no difference when one exists).

To compute the Scheffe in *R*, you will first need to load and require the *agricolae* package. Once the package is installed and the **aov()** function has run, the *R* code listed as follows will produce the output shown in Figure 6.14:

```
> scheffe.test(a1,"hospital$position", group=F,
console = T)
```

The *a1* vector and the categorical grouping variable *position* are included in the statement:

Study: a1 ~ "hospital$position"

Scheffe Test for hospital$sensitivity

Mean Square Error  : 7.378911

hospital$position,  means

```
      hospital.sensitivity        std   r Min Max
MD              13.41176 3.083400 17   9  19
PA              13.87500 2.156386 16   9  17
RN              16.05882 2.794427 17   9  19
```

alpha: 0.05 ; Df Error: 47
Critical Value of F: 3.195056

Harmonic Mean of Cell Sizes  16.65306
Comparison between treatments means

```
          Difference pvalue sig      LCL         UCL
MD - PA -0.4632353 0.8873     -2.590692  1.66422167
MD - RN -2.6470588 0.0241  *  -4.742034 -0.55208403
PA - RN -2.1838235 0.0802  .  -4.311280 -0.05636657
```

FIGURE 6.14. Scheffe output from *agricolae* package.

Signif. codes:
0 '***' 0.001 '**' 0.01 '*' 0.05 '.' 0.1 ' ' 1

The top part of the output displays information similar to what we saw in the **describeBy()** function. The *sig* column in the bottom portion of the illustration shows symbols displayed in *R* affiliated with significance codes, as displayed directly under Figure 6.14. Since the MD-RN pair shows "*", there is a significant difference at the 0.05 level, and since the PA-RN pair shows ".", there is a significant difference at the 0.1 level. Since we set our significance level at 0.05, the Scheffe confirms the previous results of the Tukey HSD: RNs are more sensitive than MDs, but not more sensitive than PAs.

## CONCLUSION

In this chapter, we discussed how to compare means. At first, we considered the simplest situation in which a sample mean was compared against a known population parameter in a one-sample *t*-test; however, very often, this is not known, and we want to know with some level of confidence where the actual population parameter lies. Confidence intervals are very helpful in making decisions and better understanding your analysis.

Very often, we might want to compare the means of two or more samples to each other. When there are only two samples, we can use the *t*-test; however, with three or more samples, we want to use ANOVA to reduce our chances of making a Type I error. Since ANOVA only tells us if there are statistically different groups between samples, post hoc testing can be used to better understand where those differences lie.

A third concept we discussed in this chapter was the notion of dependent samples. This takes place most often when subjects are observed twice and compared over time. In these cases, dependent samples *t*-tests can be used to look at mean change from the first observation to the second. Finally, measures of effect size can be used to both quantitatively describe how much change is observed and qualitatively denote clinical or practical significance (i.e., what we can actually notice), compared to measures of statistical significance, which may not be able to detect noticeable changes.

## EXERCISES

For each of the following, please use the *student* data, located in Appendix D. While we acknowledge that this sample is really too small to do parametric testing, we purposely created a small dataset for exercise purposes only.

1. The average college student takes five courses each semester. Test the hypothesis that students at Education State University are different from other college students. Can you reject the null at the 0.05 level? At the 0.01 level?
2. Construct a 95% CI to determine what the population mean is for GPA based on our sample. What is the 99% CI? Why would the 99% CI be larger than the 95% CI?
3. Test the hypothesis that men and women have different GPAs. Can you make use of the pooled standard deviation in your significance test? Why or why not?
4. Education State decided to determine how fast students read at orientation (WPM1) and again after one semester (WPM2). How would you test the hypothesis that students' speed of reading changes as they progress through college? Test that hypothesis and then construct a 95% CI to determine what the

true difference in means is. Finally, calculate an effect size and describe the degree of change from one time point to the next.

5. How would you test the hypothesis that how students perceive their success is related to their actual GPAs? Test that hypothesis. Based on your findings, do you think students are good judges of themselves as students?

## *R* EXERCISES

For the following questions, please use the *hospital* dataset that is included with this book.

1. The average length of stay for hospital patients in the United States is 5.4 days. Test the hypothesis at the 0.05 level that the patients at County General are different from other patients.

2. Consider whether patients' LOS is related to their readmission within 30 days. What version of the *t*-test (equal variances or unequal variances) would you use and why? Is there a significant difference in LOS between these two samples?

3. What is the mean ADL for patients for each of the three types of discharge plans? Is there a relationship between patients' ADLs and their discharge plans? If there is a significant relationship, where do those relationships lie? What conclusions can we draw about ADLs and discharge plans?

4. Is there a significant relationship between patients' perceptions about the sensitivity of the care they received and their overall satisfaction with their hospital stays? What is the 95% confidence interval for the difference in the means for these two samples? Do the findings of this analysis surprise you in some way?

# /// 7 /// NONPARAMETRIC TESTS OF CENTRAL TENDENCY

To work through the examples in this chapter, you will need to install and load the following *R* packages:

- *boot* (boot is installed with *R*)
- *psych*

You will also need to load the following dataset:

- *Workers.Rdata*

For more information on how to do this, refer to Appendix D.

## INTRODUCTION

In this chapter, we discuss nonparametric tests of central tendency. These are also referred to as distribution-free statistics since they do not necessitate that the data are normally distributed and are not based on probability distributions described in previous chapters. As a result of having fewer assumptions, nonparametric tests are thought to be less robust than parametric ones since it is more likely that one may decide to accept the null hypothesis when the alternative is true (i.e., make a Type II error). However, nonparametric statistics should be considered as an alternative when data are ordinal, when the data are ranked, when the *n* is small (below 50), or when there are outliers. As mentioned earlier, an important criterion of parametric tests is that the dependent variable is normally distributed. This rule can be relaxed if the sample size is 50 or greater; however, if these conditions are not met, nonparametric statistical tests will be necessary, and these are discussed in this chapter.

In this chapter, we will cover three of the most widely used nonparametric tests: the Mann-Whitney U test, the Wilcoxon signed rank test, and the Kruskal-Wallis test. Calculating these tests by hand can be extremely laborious; as a result, we provide the formulae for these in this chapter, but we will do all calculations using *R*.

## MANN-WHITNEY U TEST/WILCOXON TEST

The Mann-Whitney U test is also known as the Wilcoxon test (not to be confused with the Wilcoxon signed rank test, to be discussed in the next section) and is a nonparametric alternative to the *t*-test. It is assumed that the data are at least ordinal (i.e., not nominal) and the groups or samples being compared are independent of one another. The idea behind the test is to rank the observations in each group from lowest to highest and compare how different the sums of the ranks are from each other. If a difference exists between samples, then most of the low ranks will belong to one sample and most of the high ranks to the other sample. On the other hand, if no difference exists, then low and high ranks will be equally distributed between samples. The formula for the Mann-Whitney U test is presented in Formula 7.1. The formula is a version of *U* that approximates the normal distribution:

$$\text{Formula 7.1:} \quad U = \frac{R_1 - R_2 - (N_1 - N_2)(N+1)/2}{N_1 + N_2 \ (N+1)/3}$$

where:

$R_1$ and $R_2$ are the sums of the ranks of the separate samples,
$N_1$ and $N_2$ are the sample sizes of the separate samples, and
$N$ is the total sample size.

To illustrate this with an example, open the *worker.Rdata* file as described in Appendix D.

We can begin by looking at the variable *commun*, which is a measure of the degree of satisfaction with communication about policies and procedures at the Funky Fun Corporation. Higher scores reflect higher levels of satisfaction. The question to be tested is, "Do employees who have thought about leaving their jobs in the last year have lower levels of satisfaction with communication than those who did not think of leaving?" This information on thinking of leaving is contained in the variable *i1*.

Hypothesis testing is a bit different with this type of test. Because we are not making use of a known distribution, we simply can speculate the following:

$H_0$ = both samples (those thinking about leaving and those not thinking about leaving) are from the same population so that observations will be distributed similarly

H$_1$ = both samples are from different populations so each will be distributed differently

Notice here that there is no discussion about a difference in means or standard deviations, and this is reflected in the formula for this test.

Because the dataset contains only 30 observations, the Mann-Whitney U is considered to test this question instead of a *t*-test, since *i1* has only two levels: *Yes* and *No*.

Let's begin by describing the differences in medians between the two groups by utilizing the *psych* package's **describeBy()** function. Load the *psych* package and type the following syntax:

```
>describeBy(workers$commun,workers$i1)
```

The syntax produces the output displayed in Figure 7.1.

```
Descriptive statistics by group
group: No
    vars  n  mean   sd median trimmed  mad min max range skew kurtosis   se
X1     1 11 12.18 2.36     12   12.11 2.97   9  16     7 0.09    -1.45 0.71
-----------------------------------------------------------------------
group: Yes
    vars  n mean   sd median trimmed  mad min max range  skew kurtosis  se
X1     1 15 9.87 2.33     10      10 2.97   5  13     8 -0.33    -0.82 0.6
```

FIGURE 7.1. Output from **describeBy()** function.

The employees thinking of leaving have a median score of 10 on satisfaction with communication as compared to a median of 12 for those employees who are not thinking of leaving.

The following command will run the Mann-Whitney U in *R*, and the output is displayed in Figure 7.2:

```
>wilcox.test(workers$commun~workers$i1)
```

```
        Wilcoxon rank sum test with continuity correction

data:  commun by i1
W = 123.5, p-value = 0.03418
alternative hypothesis: true location shift is not equal to 0

Warning message:
In wilcox.test.default(x = c(11, 9, 10, 14, 12, 12, 9, 15, 16, 12,  :
    cannot compute exact p-value with ties
```

FIGURE 7.2. Output from Mann-Whitney U test/Wilcoxon test.

As you can see, $R$ produced a warning in addition to the calculated test statistic and $p$-value. The reason for this message is that there are two identical ranks for communications in the groups. When ranks are not unique, an exact $p$-value cannot be calculated.

Notice that the test statistic produced was $W$, and with $p < 0.05$, we can reject the null hypothesis and accept the alternate. The results from the Mann-Whitney U test/ Wilcoxon test indicate that workers thinking of leaving have a significantly lower median rank score of satisfaction with communication than employees who did not think of leaving within the last year.

## WILCOXON SIGNED RANK TEST

The Wilcoxon signed rank test is a nonparametric alternative to the dependent samples $t$-test. It is assumed that the data are at least ordinal and the same condition is being measured before and after an intervention on the same subjects. You could also have other types of matched pairs such as husband and wife, whose measures would not be independent of one another. The idea behind the test, for example, is to rank the observations before and after an intervention and compare how different the sums of the ranks are from each other. If a difference exists between the time periods, then most of the low ranks will belong to one time period and most of the high ranks to the other time period. On the other hand, if no difference exists, then low and high ranks will be equally distributed between time periods. The formula for the Wilcoxon signed rank test is presented in Formula 7.2:

$$\text{Formula 7.2}: \quad W = \frac{T - \dfrac{N(N+1)}{4}}{\sqrt{\dfrac{N(N+1)(2N+1)}{24}}}$$

The terms in the formula require some explanations:

$T$ = sum of the ranks of the difference in pairs
$N$ = number of pairs

As an example, let's consider that the management of the Funky Fun Corporation wants to be sure their employees are clear on what they are doing in their jobs. They measure *ambiguity*, the degree of vagueness about what is expected of employees at the Funky Fun Corporation before and after an intervention designed to improve administrative leadership. Higher scores on the ambiguity scale mean that employees think their job expectations are clearer than for those with lower scores. The question to be tested is, "Did employees' levels of ambiguity change after the intervention?"

This concept is measured by *ambiguity* before the intervention and by *ambiguity2* afterward.

Let's take a look at the medians to obtain an idea of the observed change. The following code will produce the medians:

```
median(workers$ambiguity, na.rm=T), and
median(workers$ambiguity, na.rm=T).
```

The output displayed in the console will produce the following medians: 3.625 and 3.5.

What we observe is that there was a small decrease in the medians from the first time point to the second.

The following syntax will run the Wilcoxon signed rank test in *R* and the results can be found in Figure 7.3:

```
> wilcox.test(workers$ambiguity,workers$ambiguity2,
paired=TRUE)
```

```
        Wilcoxon signed rank test with continuity correction

data:  workers$ambiguity and workers$ambiguity2
V = 125.5, p-value = 0.7141
alternative hypothesis: true location shift is not equal to 0

Warning messages:
1: In wilcox.test.default(workers$ambiguity, workers$ambiguity2, paired = TRUE)
  :
    cannot compute exact p-value with ties
2: In wilcox.test.default(workers$ambiguity, workers$ambiguity2, paired = TRUE)
  :
    cannot compute exact p-value with zeroes
```

FIGURE 7.3. Output from Wilcoxon signed rank test.

Notice that the function in both this test and the Mann-Whitney U is the same; however, the test statistic that is presented is different, and in this case, it is *V*. The *p*-value is greater than 0.05. As a result, the null hypothesis of no difference in perceptions of ambiguity before and after the intervention cannot be rejected.

## KRUSKALL-WALLIS TEST

The Kruskal-Wallis Test is a popular nonparametric alternative to the one-way analysis of variance (ANOVA) discussed in the previous chapter. The test compares the median ranks or ordinal data between two or more independent groups. The calculation is presented as Formula 7.3:

$$\text{Formula 7.3}: \quad H = \frac{\left( \dfrac{12}{N(N+1)} \sum_{i=1}^{k} \dfrac{R_i^2}{N_i} \right) - 3(N+1)}{1 - \sum T_i / (N^3 - N)}$$

The terms in the formula require some explanations:

$N_i$ = number of observations in the $i$th category

$T_i = t_1^3 - t_1$,

where $T_i$ = the number of ties in the $i$th category

As an example, let's consider that the Funky Fun Corporation is interested in understanding their employees' job satisfaction, and they wonder if the job title (*position*) is related to overall job satisfaction (*totjobsat*). Let's examine the medians of satisfaction between groups:

```
>describeBy(workers$totjobsat,workers$position)
```

The output from this function is displayed in Figure 7.4.

```
 Descriptive statistics by group
group: Clerical
    vars n mean    sd median trimmed  mad min max range skew kurtosis   se
X1    1 5 92.4 11.08     90   92.4 5.93  80 110    30 0.51    -1.41 4.96
-----------------------------------------------------------------------
group: Manager
    vars n  mean    sd median trimmed  mad min max range skew kurtosis   se
X1    1 8 102.38 11.07    109 102.38 3.71  86 112    26 -0.5    -1.83 3.91
-----------------------------------------------------------------------
group: Worker
    vars n  mean   sd median trimmed  mad min max range skew kurtosis   se
X1    1 6 97.83 9.93     96  97.83 11.12  86 112    26 0.21    -1.84 4.05
```

FIGURE 7.4. Descriptive statistics of job satisfaction by position.

The job satisfaction scale ranges from a low of 80 to a high of 112. Managers have the highest median scores (109) as compared to 96 for workers and 90 for clerical employees.

We can see if we are able to reject the null hypothesis that position is unrelated to overall job satisfaction by using the following function in *R*. The results are displayed in Figure 7.5. The following command will run the test:

```
>kruskal.test(workers$totjobsat,as.
factor(workers$position))
```

Note that the grouping variable must be defined in *R* as a factor variable. The safest approach is to always use the **as.factor()** function to ensure that the grouping variable is seen as a factor. For an in-depth discussion of factor variables see Auerbach and Zeitlin (2015).

```
        Kruskal-Wallis rank sum test

data:  workers$totjobsat and as.factor(workers$position)
Kruskal-Wallis chi-squared = 1.9968, df = 2, p-value = 0.3685
```
FIGURE 7.5. Output from Kruskal-Wallis test.

Although we observed differences between group medians, because *p* is greater than 0.05, we cannot reject the null hypothesis.

## BOOTSTRAPPING

When data are not normally distributed or there are outliers, confidence intervals and hypothesis testing become questionable. In the previous section you were introduced to nonparametric statistics, which could be considered as an alternative when data are not normally distributed.

Bootstrapping is another nonparametric method that can be utilized when the distribution of data is questionable. Crawley (2012) provides an in-depth explanation of the procedure.

Consider that you have a single sample of size *n*. You can sample that sample, replace what you have sampled, and continue to sample with replacement many times, say, 1,000 times or more. Every time you take a sample, a statistic, such as the mean or median, is calculated. Finally, after all the samples are drawn, the standard deviation and confidence intervals between samples of the statistic are calculated. This idea is like calculating confidence intervals for the *t*- or *F*-distribution, and the overall concept is similar to the KHAN ACADEMY® example shown in the video titled "Sampling distribution of the sample mean."

Let's start by looking at an example of the median in which bootstrapping is the only way to obtain a confidence interval. Luckily when you installed *R*, the package *boot* was also included; however, you do need to require it before proceeding. This package will automatically perform the sampling necessary to bootstrap correctly. It is a complicated package to run, so we have included some *R* scripts for calculating confidence intervals for several relevant statistics. Here you will need to load the script entitled *median.R*. As shown in Figure 7.6, the script file is opened in the top left pane in *RStudio*.

```
1  require(boot)
2  b.median<-workers$age  #change wroker$age to target variable
3  bootmedian<-function(b.median,i) {median(b.median[i],na.rm=T)}
4  results<-boot(data=b.median, statistic=bootmedian, R=1000)
5  results
6  plot(results)
7  boot.ci(results, type="bca")
8
```

FIGURE 7.6. *R* bootstrap script for medians.

With the *workers* dataset open, you are ready to run the script by highlighting lines 1 through 7 and clicking *Run* on the right of the tool bar. This script can be used on any appropriate data by changing the target variable name. In the syntax in Figure 7.6, *R* = represents the number of samples obtained. We have found that 1,000 is usually an adequate number. In this example, the 95% confidence interval is calculated for the median. The statistic can be changed to a different one such as the mean.

As shown in Figure 7.7, the statistical output is presented in the Console. Since the samples are drawn at random, it is likely you will have slightly different results than those illustrated here.

```
Bootstrap Statistics :
    original  bias     std. error
t1*       37 -0.4215     2.93636
> plot(results)
> boot.ci(results, type="bca")
BOOTSTRAP CONFIDENCE INTERVAL CALCULATIONS
Based on 1000 bootstrap replicates

CALL :
boot.ci(boot.out = results, type = "bca")

Intervals :
Level       BCa
95%   (32.85, 42.18 )
Calculations and Intervals on Original Scale
```

FIGURE 7.7 Example of bootstrap statistical output for a median.

The results display the total sample median of 37; the *bias* is the amount of difference between the bootstrap statistic and the one for the sample. The *std. error* is the average amount of deviation of the median value between samples. Finally, the *BCa 95% confidence interval* of 32.85 to 42.18 is also displayed. The *BCa* is a bias correction in that the method corrects for skewness in the data. It is nonparametric because it uses percentiles, the 5th and 95th percentiles, that do not require a normal distribution as the upper and lower limits of the confidence intervals.

Figure 7.8 displays the graphical output that appears in the lower right Plots pane. It displays the extremes of the possible median values produced by the bootstrapped samples. The middle point of the x-axis of the quantiles plot is the median. The integers represent deviation above and below the median. If there were no extreme values, every dot would be positioned on the dashed line.

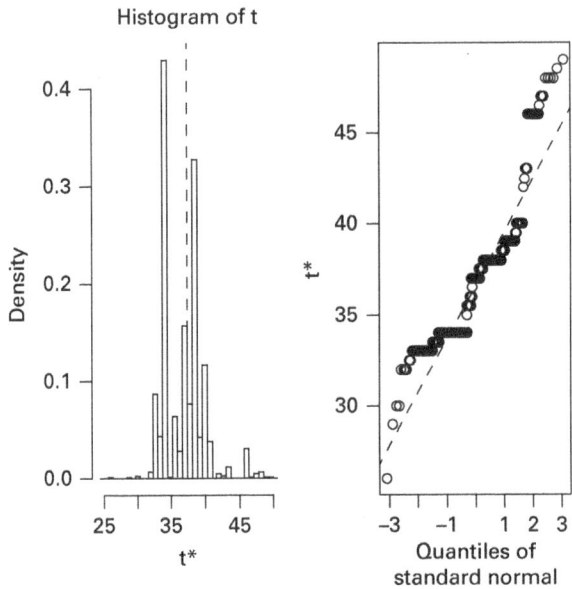

FIGURE 7.8. Graphical output from median bootstrapping script.

For an example of hypothesis testing using bootstrapping, let's look at the variable *commun*. This variable represents the degree of satisfaction with communication about policies and procedures at the Funky Fun Corporation. Higher scores reflect higher levels of satisfaction. The question to be tested is, "Do employees who have thought about leaving in the last year have lower levels of satisfaction with communication than those who did not think of leaving?" This information on thinking about leaving is, again, contained in the variable *il*. We could run a *t*-test, but the sample size is small and *commun* may not be normally distributed.

The code that follows produced the *t*-test output in Figure 7.9:

```
>t.test (workers$commun~workers$il)
```

```
       Welch Two Sample t-test

data:  commun by i1
t = -2.4872, df = 21.527, p-value = 0.02113
alternative hypothesis: true difference in means is not equal to 0
95 percent confidence interval:
 -4.2480096 -0.3822934
sample estimates:
mean in group 1 mean in group 2
      9.866667        12.181818
```

FIGURE 7.9. *t*-Test of satisfaction with communication by thought of leaving within the year.

Since our sample is small, we may want to use bootstrapping to obtain a bias-corrected confidence interval. Open the script *bootstrap of t.R.* and run as was described in the previous example. Figure 7.10 displays the output shown in the top left pane. Once again, since the samples are drawn at random, it is likely you will have a different set of samples, resulting in a marginally different standard error.

```
Bootstrap Statistics :
      original      bias      std. error
t1* -2.487226 -0.148134     1.045514
> plot(results)
> boot.ci(results, type="bca")
BOOTSTRAP CONFIDENCE INTERVAL CALCULATIONS
Based on 1000 bootstrap replicates

CALL :
boot.ci(boot.out = results, type = "bca")

Intervals :
Level          BCa
95%    (-4.49, -0.31 )
Calculations and Intervals on Original Scale
```

FIGURE 7.10. Example of bootstrap statistical output for *t*-statistic.

The output shows that the bias is small, and the standard error is slightly over 1. The bounds of the 95% confidence interval of the *t*-value are between -4.49 and -0.31. The results confirm what we observed from the *t*-test that the differences in means are not equal to zero. Figure 7.11 displays the graphical output from this script. The histogram and quantile graph display that the samples form a normal distribution.

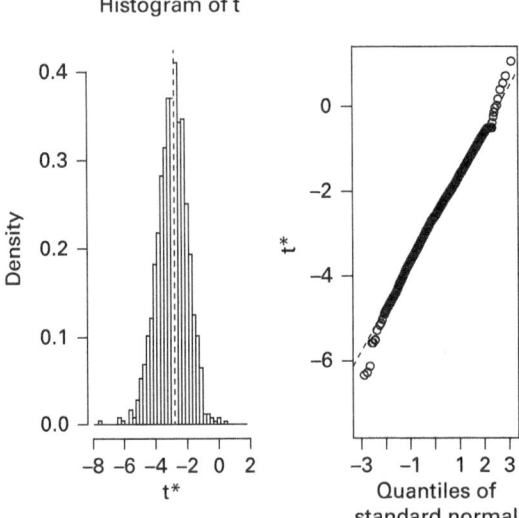

FIGURE 7.11. Graphical output from bootstrap of *t*-statistic.

A number of other scripts are available on the website for this text to bootstrap Pearson's *r*, which we will discuss in the next chapter, and the *F*-statistic and the mean.

## EXERCISES

Load the *worker.Rdata* dataset to answer each of the following questions:

1. Run the appropriate nonparametric test to answer the following question: "Do employees who have thought about leaving their jobs in the last year have lower overall satisfaction than those who did not think of leaving?" The variable *i1* is the grouping variable and *totjobsat* is a measure of overall satisfaction. Explain why you used the test selected. What are your findings? Produce the medians between groups and describe what you observe.
2. Use the median script to bootstrap the 95% confidence interval for *totjobsat* (overall satisfaction) by making the appropriate change to the *R* code. What additional information does this provide for your decision making?
3. Run the appropriate nonparametric test to answer the following question: "Did employees' levels of satisfaction with communication change after the intervention?" This concept is measured before the intervention with *commun* and after the intervention with *commun2*. Display the medians for each group and describe what you observe.

# //8/// CORRELATION AND SIMPLE LINEAR REGRESSION

To work through the examples in this chapter you will need to *install* and *load* the following *R* packages:

- *Hmisc*

You should also load the following dataset into the Environment pane:

- *hospital*

For more information on how to do this, refer to Appendix C.

## INTRODUCTION

So far, in terms of inferential statistics, we have considered situations in which at least one variable was categorical. Many times, however, there are situations in which both variables are continuous. When this is the case, the most appropriate way to look at relationships between variables is by correlations and/or linear regression.

In this chapter, we will look at examples of parametric and nonparametric correlations along with the most basic type of regression, ordinary least squares (OLS) regression, which is an extension of correlation.

## AN OVERVIEW OF CORRELATION

Correlation is an overused term by nonstatistical folks as it is often used to describe any type of relationship between two things. For example, we often hear things like "There's a correlation between gender and career choice." Statistically speaking, though, the term *correlation* is reserved to describe the relationship between two variables when at least one is continuous.

Correlations describe how two variables move together, and correlation coefficients can be either positive or negative. A positive correlation means that as one variable increases, the other one does as well. A negative correlation means that as one variable increases, the other decreases.

Figure 8.1 shows a perfect positive correlation for two variables, $x$ and $y$, and Figure 8.2 displays a perfect negative correlation.

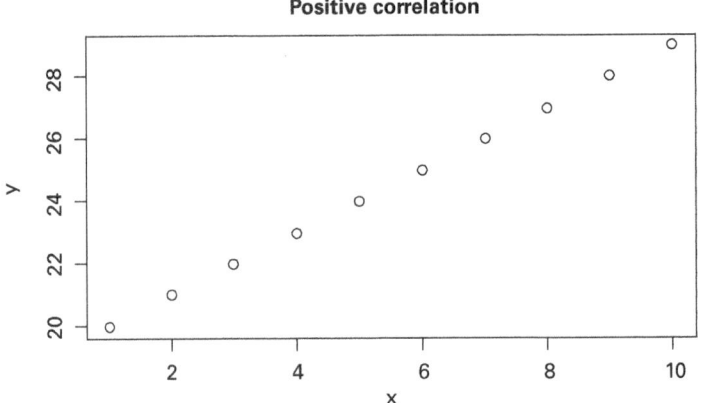

FIGURE 8.1. Positive correlation.

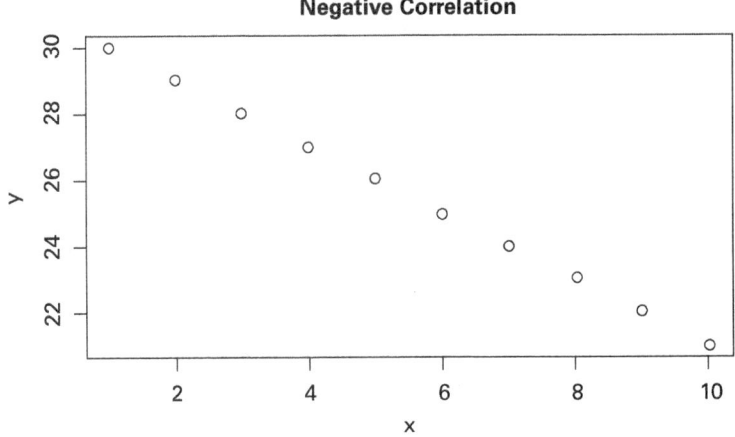

FIGURE 8.2. Negative correlation.

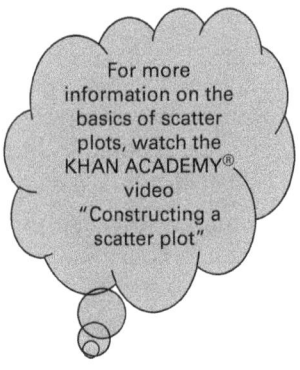

For more information on the basics of scatter plots, watch the KHAN ACADEMY® video "Constructing a scatter plot"

Notice that in the **scatter plots**, the relationship between the $x$ and $y$ variables looks linear. Scatter plots include a point for each observation denoting where the values of two variables meet.

Correlations should only be used when the relationship between two variables appears to be linear. Therefore, you should always begin looking at a relationship between two continuous variables by creating a simple plot like the ones in Figures 8.1 and 8.2.

In terms of magnitude, correlation coefficients can range from 0, indicating no relationship between the variables, to 1, with one variable perfectly predicting the second variable.

Therefore, correlation coefficients can range from −1 to +1, or, expressed mathematically, with $r$ representing the correlation coefficient:

$$-1 \le r \le +1$$

In reality, it is very unlikely that you will ever see correlations of either +1 or -1. Figures 8.3 through 8.6 show correlations of varying magnitudes and directions.

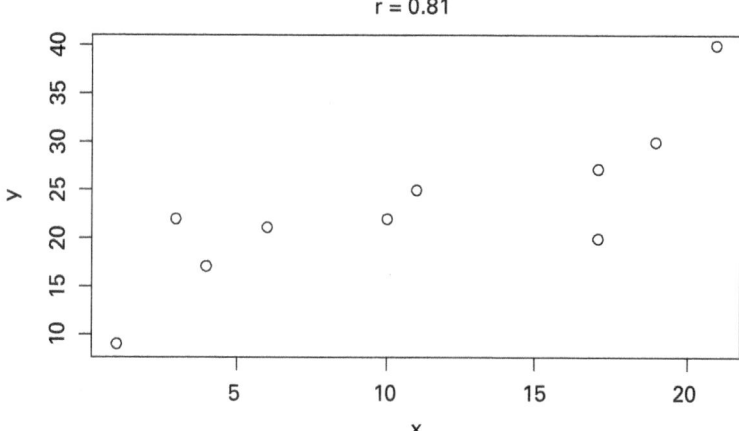

FIGURE 8.3. Strong positive correlation.

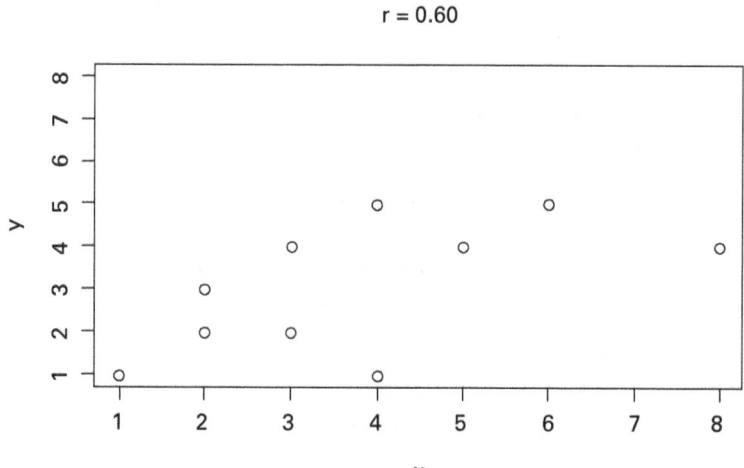

FIGURE 8.4. Moderate positive correlation.

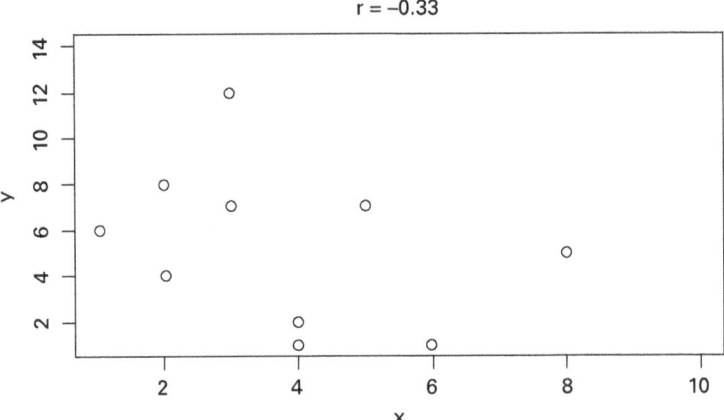

FIGURE 8.5. Weak negative relationship.

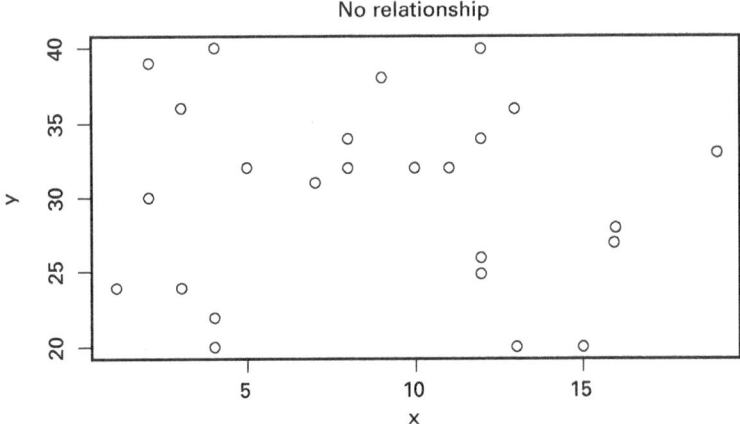

FIGURE 8.6. No relationship.

Notice that in Figure 8.6, *x* and *y* do not appear to have a linear relationship at all. Therefore, it would be inappropriate to describe the relationship between these variables in this way.

The strength of correlations can be described qualitatively, as displayed in Table 8.1. Since the strength of the relationship is not indicated by positive or negative, the descriptors in Table 8.1 are based on the absolute value of *r*.

TABLE 8.1. Interpreting Correlation Coefficients

| *r* | Description of Relationship |
| --- | --- |
| 0–0.19 | No relationship |
| 0.2–0.39 | Weak |
| 0.4–0.59 | Modest |
| 0.6–0.79 | Moderate |
| 0.8–1.0 | Strong |

## CAUTION: CORRELATION ≠ CAUSATION

It should be noted that correlations, no matter how strong, are not suggestive of causation. Simply because one variable is predictive of another is not an indicator that one causes the other.

To infer causation three conditions must be met:

1. The independent variable (the cause) must precede the dependent variable (the effect) in time.
2. The two variables must be correlated with one another.
3. The correlation between the two variables cannot be due to the influence of one or more additional variables.

While an in-depth discussion of causal relationships is beyond the scope of this book, suffice it to say that, while points 1 and 2 are relatively easy to prove, the third is not. It is very difficult—nearly impossible in fact—to discern or rule out the influence of all additional factors in determining the actual nature of the relationship between two variables. This is particularly true in the behavioral and social sciences in which environmental context, human behavior, and perceptions are often the subject of research.

Therefore, as we get further into our understanding of applied statistics, we caution you to refrain from making statements such as "Intervention X caused an improvement in school attendance" or "Depression improved as a result of Intervention Y." Statements such as these suggest causal inferences that are simply not provable in most research in the behavioral or social sciences.

> For more information on what correlations of different magnitudes look like, watch the KHAN ACADEMY® video "Example: Correlation coefficient intuition"

Alternatively, a statement like "The researchers noted a significant improvement in externalizing behavior after the introduction of Intervention Z" is more acceptable. This allows for further exploration, including the identification of additional factors that could be related to the subjects' externalizing behavior.

## CALCULATING CORRELATION COEFFICIENTS ASSUMING NORMALITY

Formulae 8.1 and 8.2 display the formulae for biased and unbiased correlations in the form of Pearson's $r$.

As stated previously, the biased formula is indicated for populations and the unbiased formula is indicated for samples:

Formula 8.1:  Biased correlation:

$$\rho_{yx} = \frac{\dfrac{\Sigma XY}{N} - \mu_X \mu_Y}{\sigma_X \sigma_Y}$$

Formula 8.2:  Unbiased correlation:

$$r_{yx} = \frac{\dfrac{1}{N-1}\left(\Sigma XY - N\,\overline{XY}\right)}{s_X s_Y}$$

In both of these equations, the numerator is a measure of the **covariance** of the variables, and the denominator is the product of the standard deviations. In the examples presented here, we will assume that the unbiased correlation coefficient is the one we want to calculate, as this will be the most appropriate formula to use in most cases.

## Calculation Example

Let's look at a problem in which you are asked to calculate a correlation. Because we are simply providing you with a calculation example, we will assume normality even though our sample size is really too small to make that assumption. Later, we will discuss how to calculate a correlation coefficient when we are unable to assume normality.

Pretend that you are working at an agency that primarily serves at-risk families, and you have been asked to determine whether there is a relationship between the number of people residing in the household and household spending on food. To do a preliminary analysis, you take a random sample of 15 clients and ask them two questions:

1.  How many people live in your household?
2.  How much did you spend last month on food that you ate at home?

The data you collect is shown in Table 8.2.

TABLE 8.2. Sample Client Data

| Client | Household Size | $ Spent on Food |
|--------|----------------|-----------------|
| 1 | 2 | 120 |
| 2 | 3 | 150 |
| 3 | 6 | 320 |
| 4 | 4 | 330 |
| 5 | 3 | 80 |
| 6 | 4 | 250 |
| 7 | 6 | 400 |
| 8 | 5 | 400 |
| 9 | 3 | 50 |
| 10 | 3 | 200 |
| 11 | 2 | 40 |
| 12 | 2 | 100 |
| 13 | 5 | 100 |
| 14 | 2 | 50 |
| 15 | 6 | 250 |

Now we need an approach for answering your research question:

**STEP 1: Determine how you are going to look at the relationship between the variables.**

As in almost all cases, you will determine this by looking at the level of measurement of each of your variables. In this case your variables are household size and money spent on food (the client column is not a variable, but is used to uniquely identify each observation). Because both of these variables are continuous, you will know that examining the relationship by calculating a correlation coefficient is appropriate.

Since you will be looking at the relationship between these variables by calculating a correlation coefficient, you will want to first create a scatter plot to determine whether it even looks like the relationship between the variables is linear.

To do this, either you will need graph paper or you can do this analysis in *R*. If you are doing this by hand, you will need to scale each variable appropriately on one of the axes. For example, since household size varies between 2 and 6, the axis on which you plot this should have these numbers. It does not matter on which axis you put each variable at this point as the overall look should be the same regardless.

To illustrate this point, look at Figures 8.7 and 8.8.

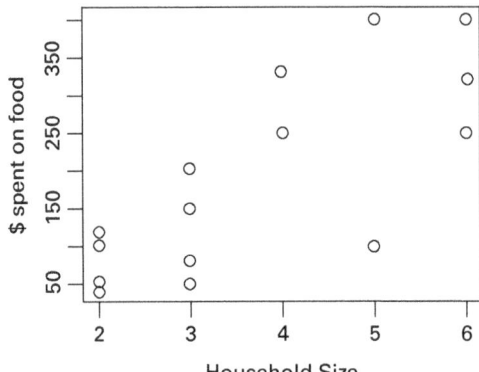

FIGURE 8.7. Scatter plot of household size and money spent on food with household size on x-axis.

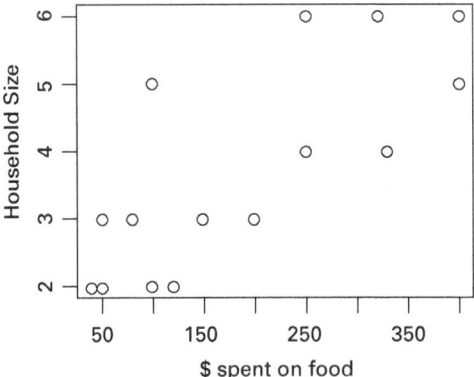

FIGURE 8.8. Scatter plot of household size and money spent on food with money spent on food on x-axis.

To do this in *R*, you could create a vector for both *x* and *y* that holds the data, as follows:

```
> size<-c(2, 3, 6, 4, 3, 4, 6, 5, 3, 3, 2, 2,
5, 2, 6)
> money<-c(120, 150, 320, 330, 80, 250, 400, 400, 50,
200, 40, 100, 100, 50, 250)
```

Notice that in entering the data, it must be entered for each variable in the exact order in which it was displayed in Table 8.2. This is because each observation is made up of both a household size and money spent on food. If we do not order this correctly, our ultimate analysis will not be accurate. To create the scatter plot displayed in Figure 8.7, simply enter the following in the Console:

```
>plot(size, money)
```

No matter which plot you look at or how you create it, it illustrates the same data. Looking at either one of these, it appears as if the relationship between the variables is, indeed, linear and sloping upward, indicating a positive relationship. Therefore, when we calculate our correlation coefficient, we know that $r$ should be a positive number.

### STEP 2: Determine which formula to use.

Your choice is whether to use the biased or unbiased formula for your correlation coefficient. Since you are obtaining information on a sample of the agency's clients (and not the entire population of clients), it is most appropriate to use the unbiased formula:

Formula 8.2:    Unbiased correlation:

$$r_{yx} = \frac{\frac{1}{N-1}\left(\sum XY - N\overline{XY}\right)}{s_X s_Y}$$

### STEP 3: Compute each of the variables in the formula.

By looking at Formula 8.2, you know you will need the following to calculate $r$:

$N$: the sample size
$\sum XY$: the sum of the products of $X$ and $Y$
$\overline{X}$ : the mean of $X$
$\overline{Y}$: the mean of $Y$
$S_x$: the standard deviation of $X$
$S_y$: the standard deviation of $Y$

Once you have these values, you can plug them into the previous formula and, using the correct order of operations, calculate your correlation coefficient.

Notice that you must assign one of your variables to be $X$ and the other $Y$. It does not matter which is which so let's say that household size is $X$ and money spent on food is $Y$.

We also know that $N = 15$ since you obtained information from 15 clients.

Now we can calculate $\sum XY$ using Table 8.3, which is taken from Table 8.2, presented in the problem itself.

TABLE 8.3. Calculating ΣXY for Sample Client Data

| Household Size (X) | $ Spent on Food (Y) | XY |
|---|---|---|
| 2 | 120 | 240 |
| 3 | 150 | 450 |
| 6 | 320 | 1920 |
| 4 | 330 | 1320 |
| 3 | 80 | 240 |
| 4 | 250 | 1000 |
| 6 | 400 | 2400 |
| 5 | 400 | 2000 |
| 3 | 50 | 150 |
| 3 | 200 | 600 |
| 2 | 40 | 80 |
| 2 | 100 | 200 |
| 5 | 100 | 500 |
| 2 | 50 | 100 |
| 6 | 250 | 1500 |

$$\Sigma XY = 12{,}700$$

Now we can compute the means of $X$ and $Y$ as we had done previously and we obtain the following:

$$\bar{X} = 3.733 \quad \text{and} \quad \bar{Y} = 189.333$$

To calculate the unbiased standard deviations for $X$ and $Y$, we will use the unbiased formula found in Chapter 2:

$$s_x = 1.534 \text{ and } s_y = 127.754$$

Now, simply plug the calculated into the formula and simplify:

$$r_{yx} = \frac{\dfrac{1}{15-1}\left[12{,}700 - (15(3.733)(189.333))\right]}{(1.534)(127.754)}$$

$$r_{yx} = \frac{\dfrac{1}{14}\left[12{,}700 - 1{,}601.701\right]}{195.975}$$

$$r_{yx} = \frac{\dfrac{1}{14}(2098.299)}{195.975}$$

$$r_{yx} = \frac{149.879}{195.975}$$

$$r_{yx} = 0.765$$

On the face of it, this manually calculated result seems correct. First, $-1 \leq 0.765 \leq +1$, which is a requirement for a correlation coefficient. Second, we knew that our coefficient had to be positive based on the scatter plots we produced earlier. Finally, the scatter plots we produced LOOKED like there was a moderately strong relationship between the two variables. Therefore, we can look at our hand-calculated correlation coefficient and get a sense as to whether our calculated value looks to be a reasonable result or not.

Interpreting our findings, it appears that as household size increases, spending on food also increases, and there is a moderately strong relationship between household size and spending on food.

## TESTING FOR TYPE I ERROR WITH PEARSON'S *R*

Hypothesis testing utilizing Pearson's $r$ is appropriate when we can assume that both variables are normally distributed and when the pairs of observations are independent of one another.

When looking at the significance of a correlation coefficient, the hypothesis being tested is:

$$H_0: \rho = 0$$

$$H_1: \rho \neq 0$$

Testing for significance uses the $t$-distribution, and $df = n - 2$, with $n$ = number of pairs in the sample. Statisticians have compiled critical values of $r$, above which $H_0$ can be rejected and $H_1$ can be accepted. Refer to Table 4 in Appendix F for a table of critical $r$ values.

In our previous example, we could ask ourselves, "Is the calculated correlation coefficient of 0.765 significantly different from 0 based on our sample size of 15 pairs?" Here, $df = 13$, and, if we set $\propto = 0.05$, we can use Table 4 to determine that $r_{crit} = 0.514$ for our two-tailed test. Since $r_{yx} > r_{crit}$, we can reject the null hypothesis that $\rho = 0$ and accept the alternate hypothesis.

## CALCULATING CORRELATION COEFFICIENTS
## WHEN WE CAN'T ASSUME NORMALITY

When we are unable to assume normality or we have few pairs of observations, Pearson's $r$ is not an appropriate method for computing a correlation coefficient. In these cases, the Spearman rank-order correlation coefficient, or Spearman's *rho*, is more appropriate.

When calculating Spearman's *rho*, each observation for each variable is ranked in order from lowest to highest. Therefore, if there are 20 observations, the ranking for each variable will go from 1 to 20. These ranks then replace the values initially assigned to each observation of each variable.

The formula for Spearman's *rho* is illustrated in Formula 8.3:

$$\text{Formula 8.3:} \quad r_s = 1 - \frac{6 \sum D^2}{N(N^2 - 1)},$$

where:

D = difference between a pair of ranks

N = number of pairs

### Dealing with Tied Ranks and a Calculation Example

Very often you will have variables in which multiple observations of the same variable all have the same value so ranks will be tied. Let's, for instance, look at the data initially presented in Table 8.2, but with columns added for ranking in Table 8.4. Notice that the smallest household size, which would be assigned a rank of "1," occurs in Clients 1, 11, 12, and 14. Since all of these cannot receive this rank, they all will receive the same rank, which will be the average of 1, 2, 3, and 4, which is 2.5. Households with three people begin with a rank of 5. Clients 2, 5, 9, and 10 should receive the ranks of 5 through 8, so each will receive the average of those, which is 6.5. Clients 4 and 6 then receive ranks of 9.5 each, and clients 8 and 13 each receive ranks of 11.5. Finally, households with six family members receive a rank of 14.

TABLE 8.4. Sample Client Data

| Client | Household Size | Ranked Household Size | $ Spent on Food | Ranked $ Spent on Food | D | D² |
|---|---|---|---|---|---|---|
| 1 | 2 | 2.5 | 120 | 7 | −4.5 | 20.25 |
| 2 | 3 | 6.5 | 150 | 8 | −1.5 | 2.25 |
| 3 | 6 | 14 | 320 | 12 | 2 | 4 |
| 4 | 4 | 9.5 | 330 | 13 | −3.5 | 12.25 |
| 5 | 3 | 6.5 | 80 | 4 | 2.5 | 6.25 |
| 6 | 4 | 9.5 | 250 | 10.5 | −1 | 1 |
| 7 | 6 | 14 | 400 | 14.5 | −0.5 | 0.25 |
| 8 | 5 | 11.5 | 400 | 14.5 | −3 | 9 |
| 9 | 3 | 6.5 | 50 | 2.5 | 4 | 16 |
| 10 | 3 | 6.5 | 200 | 9 | −2.5 | 6.25 |
| 11 | 2 | 2.5 | 40 | 1 | 1.5 | 2.25 |
| 12 | 2 | 2.5 | 100 | 5.5 | −3 | 9 |
| 13 | 5 | 11.5 | 100 | 5.5 | 6 | 36 |
| 14 | 2 | 2.5 | 50 | 2.5 | 0 | 0 |
| 15 | 6 | 14 | 250 | 10.5 | 3.5 | 12.25 |
| | | | | | $\sum D^2$ | 137 |

Looking at money spent on food, Client 11, with spending of $40, receives the rank of 1. As illustrated in Table 8.4, some ranks are tied on this variable and others are not.

To calculate $D$ for each pair, we will simply subtract the rank for household size from the rank for money spent on food. Notice that some values will be positive and others negative, but all values for $D^2$ will, of course, be positive. Ultimately, $\sum D^2 = 137$.

Now, we can calculate Spearman's *rho*, which is really the more appropriate correlation coefficient for this dataset, given the small number of observations:

$$r_s = 1 - \frac{6\sum D^2}{N(N^2-1)} = 1 - \frac{6(137)}{15(225-1)} = 1 - \frac{822}{3,360} = 1 - 0.245 = 0.755$$

### Testing Hypotheses with Spearman's rho

The hypothesis we are testing with Spearman's rho is similar to if we were testing using Pearson's *r*:

$H_0$: ranks are independent in the population from which the sample was taken
$H_1$: ranks are not independent in the population from which the sample was taken

Because the hypothesis test is not, however, parametric, we cannot use the same table of *p*-values when the sample size is small. Therefore, if $n > 30$, we will use the same table that we used for Pearson's *r* (Table 4 in Appendix F); however, if $n \leq 30$, you will use Table 5 in Appendix F. As usual, we will set $\propto = 0.05$ for a two-tailed test. Since our calculated value $r_s (0.755) > r_{s_{crit}} (0.521)$, we can reject the null hypothesis that there is no relationship between the ranks of the pairs in the population.

### CALCULATING CORRELATION COEFFICIENTS USING *R*

For this example, we will use data from the *hospital* dataset.

As part of your ongoing study of patients at County General Hospital, your supervisor asks you to determine whether there is a relationship between patients' ages and their length of stay. Since both of these variables are continuous, you know that you need to look at the correlation between these.

As stated earlier, you will want to start by displaying the data in a scatter plot to determine if there appears to be a linear relationship between the variables. To do this, enter the following into the Console:

```
> plot(hospital$age, hospital$LOS)
```

Your output should resemble Figure 8.9.

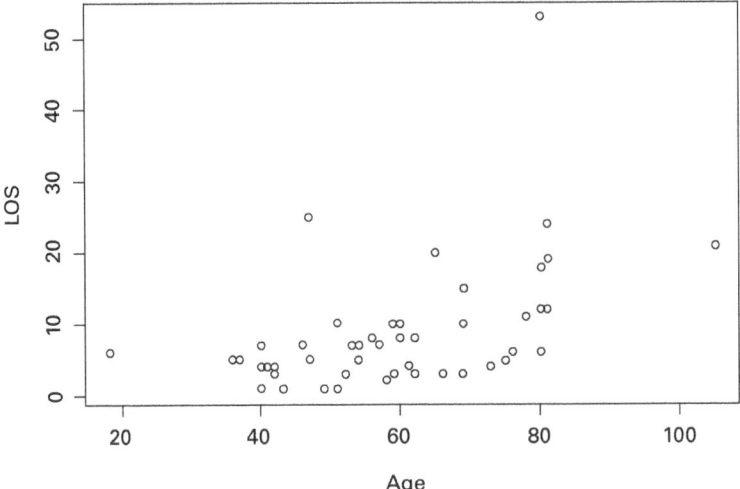

FIGURE 8.9. Scatter plot of number of patients' ages and LOS.

Notice that there does seem to be some linear relationship between these variables. The younger patients seem to have shorter lengths of stay and the older patients longer stays. There is an extreme outlier with a patient of about 80 years old with a more than 50-day hospital stay.

To actually compute the correlation coefficient, we will use a function from the *Hmisc* package:

```
>rcorr(hospital$age, hospital$LOS)
```

The output for this is found in Figure 8.10.

```
         x    y
x  1.00 0.49
y  0.49 1.00

n= 50


P
     x       y
x            3e-04
y  3e-04
```

FIGURE 8.10. Results of rcorr() function.

In the top portion of this output, you will see the correlation coefficient for the two variables. Since $x$ has a perfect correlation with $x$ (itself) and $y$ has a perfect correlation with $y$ (also itself), we will ignore the correlation coefficient of 1. We are, however, interested in the correlation coefficient looking at the relationship between $x$ and $y$, which is 0.49. In the second portion of this output, we see how many observations we have for each of our variables. Note that there is no missing data and this analysis is based on 50 pairs. Finally, the bottom section in this output is the obtained probability value. Notice that the $p$-value is written in scientific notation because $p < 0.00$. We can reject the null hypothesis and accept the alternate that the true population correlation is not equal to zero.

If you want to see the $p$-value in standard notation, enter the following into the Console:

```
> format(3e-04, scientific = F)
```

The result is shown as follows:

```
[1] "0.0003"
```

Also notice that the scatter plot and correlation coefficient seem to be in sync. Qualitatively, the correlation coefficient shows a modest positive relationship between the variables, and in the scatter plot, we notice something similar.

Now, suppose you wanted to analyze this same data using Spearman's *rho* instead of Pearson's $r$. In that case, simply add an option to the `rcorr()` function:

```
>rcorr(hospital$age, hospital$LOS, type="spearman")
```

The output for this is shown in Figure 8.11.

```
        x    y
x  1.00 0.53
y  0.53 1.00

n= 50


P
      x  y
x        0
y  0
```

FIGURE 8.11. Spearman's *rho* example.

Notice that while Pearson's *r* and Spearman's *rho* were similar for both of our examples, they were not the same. While in our examples both correlations were statistically significant regardless of the method used to calculate them, this is not always the case. Therefore, you should carefully select which correlation coefficient to use based on the characteristics of your data.

## LINEAR REGRESSION

While it is often useful to examine the relationship between two variables, it is frequently even more helpful to use what we have learned so far to predict the value of one variable given a value for another variable. Earlier in this chapter, we looked at the correlation between household size and amount of money spent on food. Using this information, could we predict how much money is spent in a household that is much larger than what we observed in the sample? The answer to this is YES!

Consider the case of a perfect correlation with *r* = 1 as illustrated in Figure 8.12.

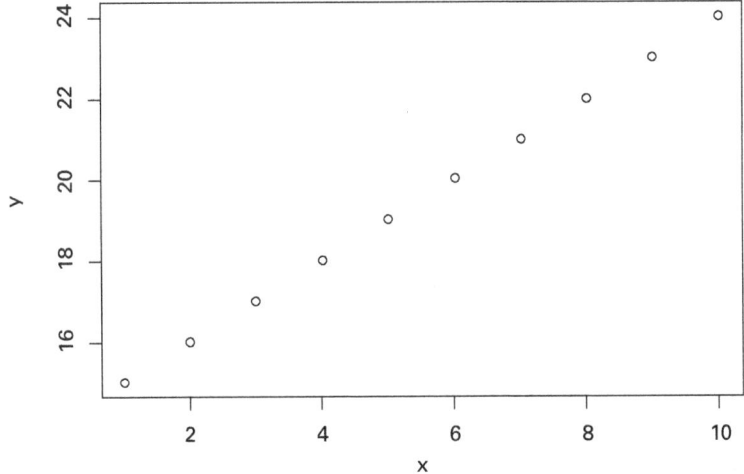

FIGURE 8.12. A perfect correlation with *n* = 10.

We could draw a straight line through these data points that could be extended infinitely in each direction. This line is the best-fitting line for this data because it goes through every point, as illustrated in Figure 8.13. This line could be used to predict future events if it fit the data well.

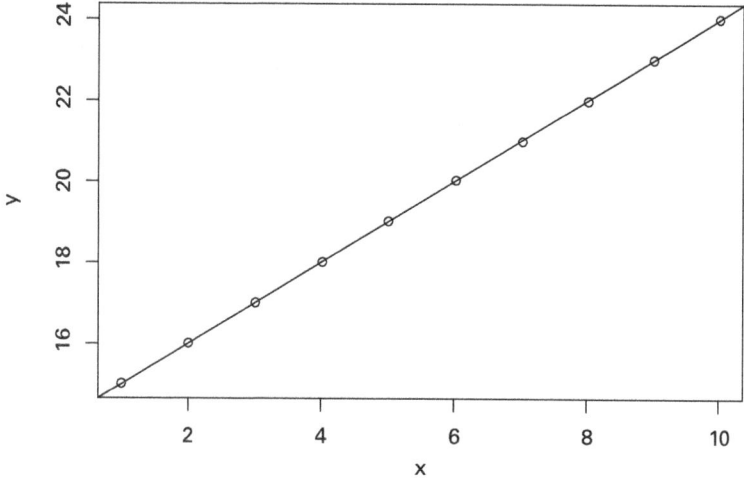

FIGURE 8.13. A perfect correlation with a line of best fit.

You may remember from math classes that every line can be defined by two things: where it crosses the y-axis and its slope, which is defined as $\frac{rise}{run}$ or $\frac{\Delta y}{\Delta x}$. In this case, we see that the y-intercept is 14 and the slope of the line is 1. Therefore, this line, which best fits the data, is defined by the equation $y = 1x + 14$ or, simplified, $y = x + 14$.

The idea with this regression line is that it can be used to predict future data points. For instance, if we know the value for $x$, we can now predict a value for $y$ even though we have never observed either $x$ or $y$.

To illustrate how this could be useful, let's consider that $x$ represents the amount of time that students study for a statistics test in minutes and $y$ represents scores on that test. With the regression line defined as $y = x + 14$, we know that if a student doesn't study at all, he or she would be expected to get only a 14 on that test. Not too good. What score could a student expect if he or she studies for an hour?

$$y = 60 + 14 = 74$$

That's better, but let's say another student, knowing this information, really wants an A on the test. Since the lowest A is a 93%, how much time should he or she expect to study?

$$93 = x + 14$$
$$x = 79 \text{ minutes}$$

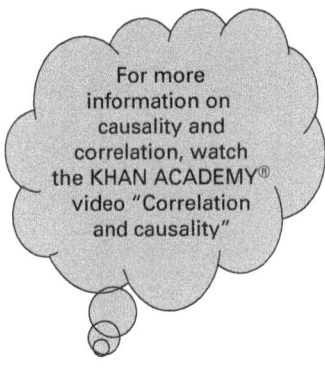

For more information on causality and correlation, watch the KHAN ACADEMY® video "Correlation and causality"

With only a slightly greater investment in time, this student is predicted to get a much better score.

The previous example is ideal because every data point is actually on the regression line, which is defined as the straight line of best fit given a set of data. But what do we mean by "best fit"?

To think about this further, consider the scatter plot displayed in Figure 8.14 that shows the relationship between patients' ages and their lengths of stay with several possible lines of best fit. Which one do you think best fits the data?

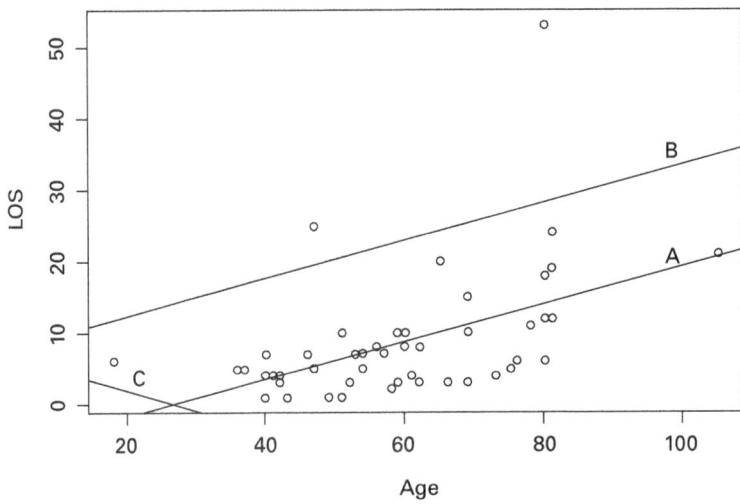

FIGURE 8.14. Scatter plot with three possible lines of best fit.

In this example, the correlation between these variables is definitely NOT perfect, as you see, with $r = 0.49$. In this instance, it is not so clear what the line of best fit would be, but you might have guessed, given your options, that the line of best fit is actually "A." So the question is, what defines the line of best fit?

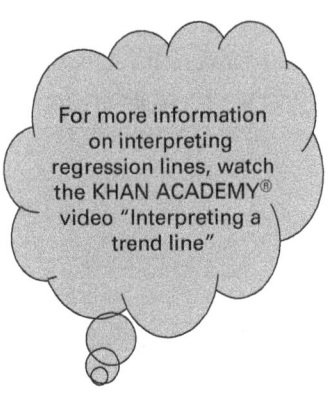

For more information on interpreting regression lines, watch the KHAN ACADEMY® video "Interpreting a trend line"

The answer of what defines the line of best fit is when the sum of squares in the definition of the line is minimized—that is, the line that is defined when the square of the **residuals**, which is the difference between predicted values (those *y*-values that would be calculated when you have the slope and y-intercept) and observed values (the values for your dependent variable for each observation), is closest to zero. Another term for this type of regression is OLS regression. To illustrate this, look at Figure 8.15.

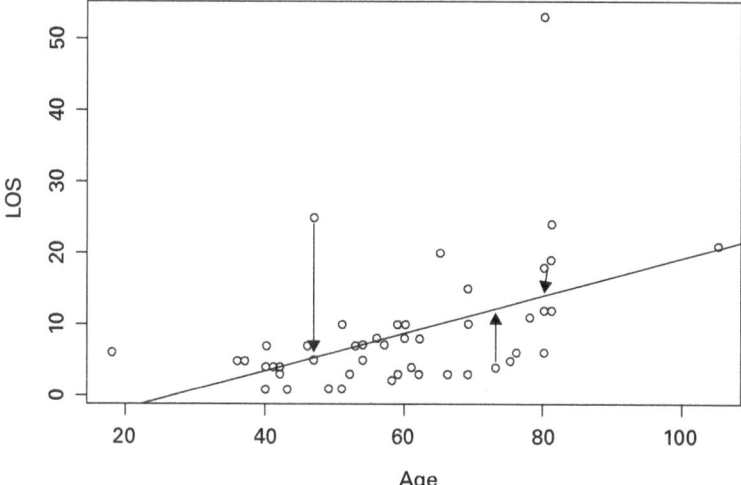

FIGURE 8.15. Minimizing the difference between observed and predicted values.

This line, which is defined by the equation $\hat{y} = -7.08 + 0.264x$, is the one that creates the shortest distance from all observations to the corresponding predicted value. The predicted value for each $x$, then, would be the one perpendicular to it on the regression line. That distance between the observed and predicted values is the residual for that observation. For instance, in Figure 8.15, consider the outlier who was 80 years old and spent 53 days in the hospital (the observed length of stay [LOS]). Using our OLS regression line, an 80-year-old person's expected LOS in the hospital would be $\hat{y} = -7.08 + 0.264x = -7.08 + 0.264(80) = 14$ days (which is where the regression line passes through). The distance between those, or the residual, as illustrated in Figure 8.15 is 53 days − 14 days = 39 days.

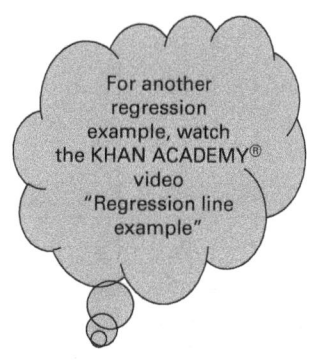

For another regression example, watch the KHAN ACADEMY® video "Regression line example"

Notice that in defining the OLS regression line, we do not use the variable $y$; rather, we use the term $\hat{y}$ (described in words as "$y$-hat") to indicate that this is a predicted value and not one that was actually observed.

## HOW TO CALCULATE THE ORDINARY LEAST SQUARES REGRESSION LINE

Now that you have an understanding of what a regression equation is, how do you figure out what that best-fitting line is? The overall formula for the regression line is shown as Formula 8.3:

$$\text{Formula 8.3:} \quad \hat{y} = b_{yx}x + a_{yx}$$

where $b_{yx}$ is the slope of the line defined by the variables $y$ and $x$ and $a_{yx}$ is the y-intercept of the line defined by the variables $y$ and $x$.

Like other formulae, the formula for the OLS regression line is similar, but not the same, for both the population, which is biased, and samples, which are unbiased. When considering the slope of the line, the biased formula is shown as Formula 8.4:

$$\text{Formula 8.4:} \quad b_{yx} = \frac{\sigma_y}{\sigma_x}\rho$$

and the biased formula for the y-intercept is shown as Formula 8.5:

$$\text{Formula 8.5:} \quad a_{yx} = \mu_y - b_{yx}\mu_x$$

When considering the slope of the line for a sample, the unbiased formula is shown as Formula 8.6:

$$\text{Formula 8.6:} \quad b_{yx} = \frac{s_y}{s_x}r$$

and the unbiased formula for the y-intercept is shown as Formula 8.7:

$$\text{Formula 8.7:} \quad a_{yx} = \bar{Y} - b_{yx}\bar{X}$$

Regardless of whether you are calculating the unbiased or biased regression line, it is imperative that you compute the slope first, as it is needed for the calculation of the y-intercept.

## Calculation Example

Refer back to the example earlier in this chapter when we wanted to look at the relationship between household size and money spent on food. In our example, the largest household size consisted of six people, but what if we wanted to project money spent on food when the household consists of seven members?

**STEP 1: Determine how you are going to look at the relationship between the variables.**

To do this, we will have to use an OLS regression since we are being asked to project something for data we do not have.

**STEP 2: Determine which formula to use.**

Like our previous example, we will need to use unbiased formulae because we are using data based on a sample and not a population. Therefore, we will use the overall formula for an OLS regression line:

$$\hat{y} = b_{yx}x + a_{yx}$$

and our calculations for the slope and y-intercept will be the unbiased versions:

$$b_{yx} = \frac{s_y}{s_x}r$$

and

$$a_{yx} = \bar{Y} - b_{yx}\bar{X}$$

This is pretty easy to compute since we did almost everything when we computed the correlation coefficient, $r$, previously. What we do need to do is to determine which is our dependent variable, $y$, and which is our independent variable, $x$. Since the question asks us what is the amount of money spent on food when the household size consists of seven people, we know that $y$ = money spent on food is the dependent variable and $x$ = household size is the independent variable. That is, we want to predict the dependent variable from the independent variable.

**STEP 3: Compute each of the variables in the formula.**

To calculate the OLS regression line, we will need the following to determine the slope using $s_y$, $s_x$, and $r$. We already have this information from our work in computing the correlation coefficient:

$s_y = \$127.754$
$s_x = 1.534$ people
$r = 0.765$

Therefore, we can calculate the unbiased slope of the regression line:

$$b_{yx} = \frac{s_y}{s_x}r = \frac{127.754}{1.534}0.765 = 63.71$$

which means that for each additional household member, there is a corresponding $63.71 increase in spending on food. Now we can calculate the y-intercept because we have everything we need to do so:

$$\bar{y} = \$189.33$$

$$\bar{x} = 3.733 \text{ people}$$

$$b_{yx} = 63.71$$

$$a_{yx} = \bar{y} - b_{yx}\bar{x} = 189.33 - (63.71)(3.733) = 189.33 - 237.83 = -48.50$$

With a negative y-intercept, we can understand that with a household size of zero, which is technically impossible, there would be negative spending on food.

Now, we can write the OLS regression model:

$$\hat{y} = b_{yx}x + a_{yx} = 63.71x - 48.50$$

Now that we have our formula for our regression model, we can predict spending on food when the household size is seven people:

$$\hat{y} = b_{yx}x + a_{yx} = 63.71(7) - 48.50 = \$397.47$$

We can visualize this by creating a scatter plot and inserting the regression line as shown in Figure 8.16.

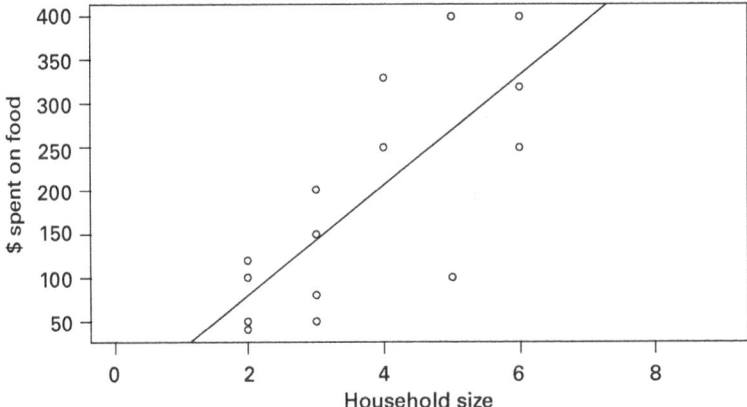

FIGURE 8.16. Scatter plot with fitted regression line.

Now, if you look at a household size of seven, you can see that the corresponding value on the y-axis is just about $400, which is what we calculated.

## HOW GOOD IS OUR MODEL?

Look back at Figure 8.11. Notice that the regression line fits the data perfectly. That is, the regression line goes through every single point and, therefore, $r = 1.0$. Now think back to the regression line fit we just calculated for predicting household spending on food based on household size. We fit the best line for the data, but how good is it? With $r = 0.765$, one variable does not perfectly predict the other, but by now you may have realized that the greater the correlation coefficient, the better the fit of the regression line.

The **coefficient of determination** is a goodness-of-fit statistic that describes how well a regression equation fits a set of data. The coefficient of determination is $r^2$ and describes the proportion of the dependent variable that is explained by the regression model. In a simple OLS regression with only one predictor (i.e., independent variable), you can calculate the coefficient of determination by squaring the correlation coefficient. The coefficient of determination ranges between 0 and 1:

$$0 \leq r^2 \leq 1$$

In Figure 8.11, $r = 1$, so $r^2 = 1$. This means that the regression model explains 100% of the variance of the dependent variable. In our previous example, $r = 0.765$, so $r^2 = (0.765)^2 = 0.585$. This means that the regression model with the single predictor, household size, explains 58.5% of the variance of spending on food.

Another commonly used goodness-of-fit statistic in OLS regression is the **standard error of the estimate**. This fit statistic measures the average deviation of predicted values from observed values. For a population, the biased formulae for this are shown in Formulae 8.8 and 8.9:

Formula 8.8 and Formula 8.9:
$$\sigma_{est} = \sqrt{\frac{\Sigma(Y - \hat{Y}Y)^2}{N}} \quad \text{or} \quad \sigma_{est} = \sigma_y\sqrt{1 - r_{yx}^2}$$

For a sample, the unbiased formulae for this are shown in Formulae 8.10 and 8.11:

Formula 8.10 and Formula 8.11:
$$s_{est} = \sqrt{\frac{\Sigma(Y - \hat{Y})^2}{N - 2}} \quad \text{or} \quad s_{est} = s_y\sqrt{\frac{N-1}{N-2}(1 - r_{yx}^2)}$$

For more information on residuals and the theory behind regression, watch the KHAN ACADEMY® videos "Introduction to residuals and least squares regression" and "Squared error of regression line"

For more information on goodness of fit, watch the KHAN ACADEMY® videos "R-squared or coefficient of determination" and "Squared error of regression line"

For another example of calculating a regression line, watch the KHAN ACADEMY® video "Calculating the equation of a regression line"

If we wanted to calculate the standard error of the estimate in our prediction of spending on food based on household member size, we could use Formula 8.11 since we already have all the inputs:

$$S_{est} = s_y \sqrt{\frac{N-1}{N-2}(1-r_{yx}^2)} = 127.754 \sqrt{\frac{15-1}{15-2}(1-0.585)}$$

$$= 127.754 \sqrt{\frac{14}{13}(0.415)} = 127.754\sqrt{0.447} = 127.754(0.669) = 85.406$$

This suggests that the average deviation of expected values from observed values is 85.406.

## COMPUTING ORDINARY LEAST SQUARES REGRESSION LINES AND GOODNESS OF FIT STATISTICS USING *R*

Using *R* to calculate the OLS regression model is easy. Let us return to our previous example in which we looked at the relationship between patients' length of stay and their ages. The first thing we will do is create a vector to store the data created by the regression. In the second step, we will view what we created in the regression. The **lm()** function means *linear model*. As usual, you can name the vector almost anything you want, but in this example, we will call this vector *r1*:

```
> r1<-lm(hospital$LOS~hospital$age)
> summary(r1)
```

The results of this are displayed in Figure 8.17.

```
Call:
lm(formula = hospital$LOS ~ hospital$age)

Residuals:
   Min     1Q Median     3Q    Max
-8.261 -4.600 -1.034  1.782 38.884
                                         ➤ y-intercept
Coefficients:                            ➤ slope
              Estimate Std. Error t value Pr(>|t|)
(Intercept)   -7.08296    4.11618  -1.721 0.091735 .
hospital$age   0.26499    0.06771   3.914 0.000286 ***
---
Signif. codes:  0 '***' 0.001 '**' 0.01 '*' 0.05 '.' 0.1 ' ' 1

Residual standard error: 7.786 on 48 degrees of freedom
Multiple R-squared:  0.2419,     Adjusted R-squared:  0.2261
F-statistic: 15.32 on 1 and 48 DF,  p-value: 0.000286
```

FIGURE 8.17. Output for OLS regression.

In the figure, we have noted which value is the y-intercept and which is the slope. With this information, we can write the formula for the regression line:

$$\hat{y} = b_{yx}x + a_{yx} = 0.26499x - 7.08296$$

The fit statistics discussed earlier, the coefficient of determination and the standard error of the estimate, are also highlighted, with the coefficient of determination being referred to as the "Multiple $R$-squared" and the standard error of the estimate being referred to as the "Residual standard error."

There is another hypothesis test related to regression models that is associated with each independent variable. In this case, we have only one and it is *age*. The *p*-value associated with this test addresses the following hypotheses:

$$H_0: b_{yx} = 0$$

$$H_1: b_{yx} \neq 0$$

When the slope of the regression line is 0, then the line of best fit is a horizontal line.

In our example, an $F$-statistic is calculated based on 1 and 48 degrees of freedom and the statistic ($F = 15.32$, $p = 0.000286$), indicating that there is only a very small chance that the contribution of the variable *age* to the overall model is zero, and, therefore, we can accept the alternate hypothesis that *age* is a significant predictor of LOS. The model we developed is superior to one without any predictors.

## Diagnostic Plots

While goodness-of-fit statistics are useful, they are not the end all and be all. $R$-squared, for example, does not estimate the influence of outliers in the coefficients or predictions. Because of this, we always recommend you assess a regression model more fully by evaluating diagnostic plots.

Most statistical packages produce diagnostic plots to help you better understand the fit of your data in regression, and $R$ is no different. After you produce the vector with the regression, we can use the `plot()` function to produce these. As an illustration, we can look at the model we developed previously by entering the following into the Console:

```
>plot(r1)
```

You will receive prompts in the Console to enable you to scroll through four diagnostic plots, displayed in the Plots pane. The first displays residuals versus fitted values. The illustration for the regression of age on LOS is displayed in Figure 8.18.

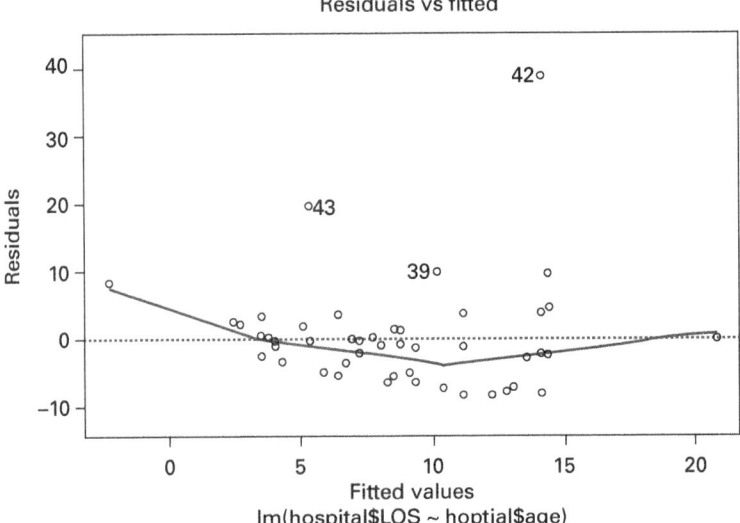

FIGURE 8.18. Residual versus fitted diagnostic plot.

The dotted horizontal line at zero denotes a perfect fit for each observation. Ideally, you would want to observe the dots to be randomly around this zero point, which indicates a linear relationship between the variables and homogeneous variance. In our case, we do not see that, and three influential cases are shown. Observation 39 is a 65-year-old female with a 20-day LOS, observation 42 is an 80-year-old male with a 53-day length of stay, and observation 43 is a 47-year-old male with a 24-day length of stay. With unusually long lengths of stay for their age, they pull this graph down.

The Normal Q-Q plot, illustrated in Figure 8.19, is used to help determine whether the data illustrated come from some theoretical sampling distribution, in this case the normal distribution. Ideally, you would like to see all points lying along the dotted line. Again, we see three observations (39, 42, and 43) that deviate from this.

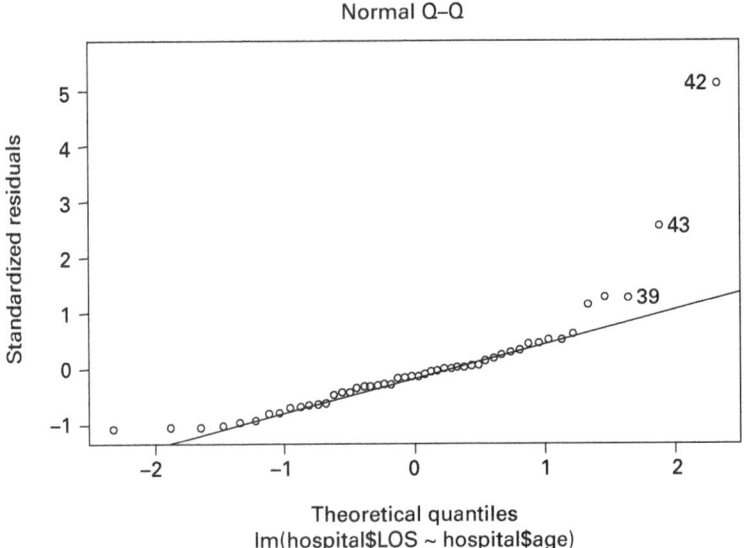

FIGURE 8.19. Q-Q diagnostic plot.

The third diagnostic plot, the Scale Location plot, is used to see if residuals are dispersed evenly among predictors. In this case, we have only one predictor (age), and this plot is displayed in Figure 8.20.

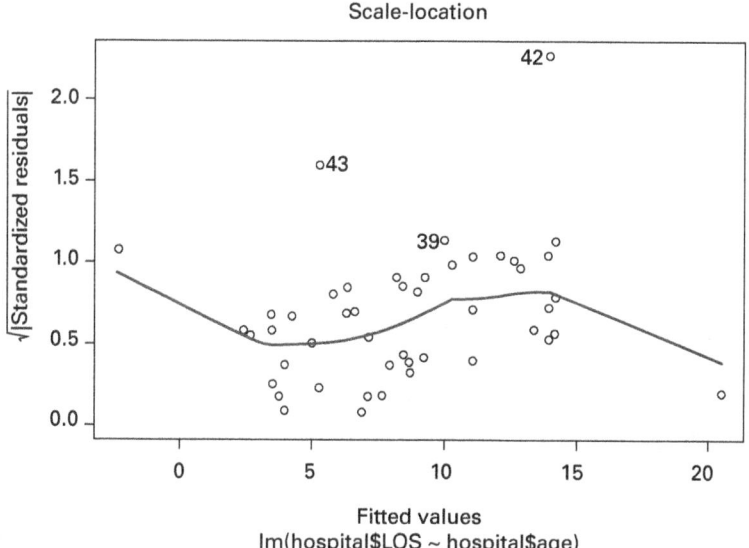

FIGURE 8.20. Scale location plot for age predicting LOS.

Ideally, we would want to see the line (which will be red when you produce it in *R*) to be fairly horizontal, which it is here. We still see our three influential cases, but observation 39 seems less problematic than observations 42 and 43.

Finally, Figure 8.21 illustrates residuals versus leverage. This plot helps identify observations that actually impact the regression model itself. Observations outside of a dotted line, in this situation case number 42, may be problematic in some way.

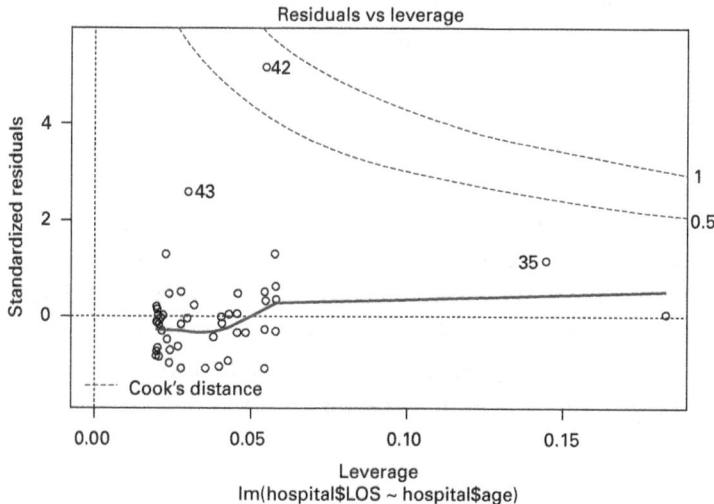

FIGURE 8.21. Residuals versus leverage diagnostic example.

**What to Do When There Is Trouble**

In our diagnostic plots, we see that we have a few observations that could be problematic to our model. When this is the case, you may want to consider a couple of things. Most simply, is there a problem with entering data? That is, was there some sort of error when you created your data file? If so, this is easily remedied by fixing it and then rerunning your regression. If that is not the case, you should look into whether the problematic observations are typical for the population you are studying. In this case, how unusual is it to have an 80-year-old with a 53-day length of stay? In our dataset, this observation is a real outlier with the next longest length of stay being only 25 days. If you look into this further and see that this is truly an atypical situation, you might consider dropping that observation from your dataset, rerunning your regression, and seeing if your fit statistics and diagnostic plots improve.

For more information on the impact of outliers on regression equations, watch the KHAN ACADEMY® video "Impact of removing outliers on regression lines"

**CONCLUSION**

In this chapter, we introduced the idea of correlation, which looks at the relationship between two continuous variables. If normality in both variables can be assumed, Pearson's *r* is the preferred method for calculating a correlation coefficient between two continuous variables. If this is not the case and a nonparametric form of correlation is needed, Spearman's *rho* is often used. The interpretation of both correlation coefficients is the same, and you were also introduced to hypothesis testing of both Pearson's *r* and Spearman's *rho*.

Later in this chapter, we examined the relationship between correlations and linear regression, which is often referred to as a trend line or the line of best fit. Since it is always possible to calculate a regression line, this chapter concluded by looking at indicators of goodness of fit. These statistics give some sense as to how well a regression line fits the data in a particular dataset. Finally, we showed you how to produce and look at diagnostic plots in *R* to look more closely at how well your data fit the derived regression model. In our example, we found how an outlier can impact an overall statistical model, and we looked at ways of remedying this.

In terms of regression, we introduced you to the simplest form in this book; however, there are other types of regression that are useful in other situations. For a more advanced discussion on multiple regression, an extension of the simple OLS regression but with more than one predictor, we refer you to *Making Your Case: Using R for Program Evaluation* (Auerbach & Zeitlin, 2015).

## EXERCISES

For these questions, use the *student* dataset provided with this book.

1. How strong of a relationship exists between students' GPA and their stress level? Create a scatter plot. What do you think? Now, calculate both Pearson's *r* and Spearman's *rho*. Are these correlation coefficients significant at the 0.10 level? At the 0.05 level?
2. How strong of a relationship exists between students' GPAs and their reading speed after intervention? Again, create a scatter plot. Does it look like there is a linear relationship between these variables? Now, calculate both Pearson's *r* and Spearman's *rho*. Are these correlations significant at either the 0.10 level or 0.05 level?
3. Looking further at the relationship between GPA and stress, calculate the formula for the regression line of how well stress predicts GPA and draw that line on your plot. Now, predict the GPA for someone with a stress level of 5.

## *R* EXERCISES

For the following questions, use the *hospital* dataset included with this book.

1. Create a scatter plot to illustrate the relationship between LOS and ADLs. Does it look like there is a linear relationship? How strong of a relationship is there between these variables? Is this relationship significant at the 0.05 level? What do these findings mean?
2. Now, develop an OLS regression model to predict LOS from ADL. What is the formula for that model? What is the predicted LOS for patients with perfect ADL scores of 12? Is ADL a significant predictor of LOS? How do you know? How much of the variance of LOS does ADL predict?
3. Finally, create the diagnostic plots for this model and include them in your homework. Do you see any problems with your model? How would you address them?

# TESTS OF ASSOCIATION

# 9

To work through the examples in this chapter, you will need to *install* and *load* the following *R* packages:

- *gmodels*
- *vcd*
- *vcdExtra*
- *rapportools*

You will also need to open the *hospital* dataset.
For more information on how to do this, refer to Appendix C.

## INTRODUCTION

In this chapter, we will discuss how to detect if there is a relationship between two categorical variables. The chi-square ($X^2$) test discussed in this chapter provides a method to compare categorical variables regardless of how many categories each contains. This goes beyond simply looking at tables and graphs, as we did earlier, as chi-square testing is a type of hypothesis testing.

There are situations in which both variables are dichotomized (i.e., they have only two categories each) and sample sizes may be small. In these cases, another test, Fisher's exact, may be more appropriate. We also address this test here, along with McNemar's test, which can only be used in very specific situations.

Finally, in this chapter, we discuss how to get additional information about the strength of the relationship between categorical variables once we establish that one exists.

## HYPOTHESIS TESTING

Since we are addressing the relationships between categorical variables in this chapter, we first need to construct both null and alternate hypotheses. In all cases, we will be testing similar hypotheses:

H₀: There is no association between the variables.
H₁: There is an association between the variables.

The way these are written is similar to the nondirectional hypotheses that we discussed earlier, as there is no statement about what we think that relationship might look like, only that one exists.

## THE CHI-SQUARE TEST

The chi-square $(X^2)$ can be used to test whether or not the frequencies obtained in a sample differ significantly from what would be expected by chance. Let's consider a very simple example to start. You and your friends wonder if all six M&M colors (red, orange, blue, brown, green, and yellow) are equally likely to end up in the bag you just bought at 7-Eleven. What we really want to know is if what we observe differs from what we expect. In this case, your hypotheses are as follows:

H₀: All colors are equally likely to be present.
H₁: All colors are not equally likely to be present.

To begin, you open the bag of M&Ms and begin sorting them by color. Your findings are displayed in Table 9.1.

TABLE 9.1. Counts of M&Ms

| | Color | | | | | | |
| --- | --- | --- | --- | --- | --- | --- | --- |
| | Red | Orange | Blue | Brown | Green | Yellow | Total |
| Observed | 12 | 18 | 22 | 11 | 14 | 13 | 90 |

Since we want to test the hypothesis related to all colors being equally likely to be present, we would expect that all 90 M&Ms would be equally distributed with each color occurring 15 times since 90 candies/6 colors = 15 candies/color. Therefore, we can expand Table 9.1 to reflect those expectations in Table 9.2.

TABLE 9.2. Observations of M&Ms Compared to What Is Expected

| | Color | | | | | | |
| --- | --- | --- | --- | --- | --- | --- | --- |
| | Red | Orange | Blue | Brown | Green | Yellow | Total |
| Observed | 12 | 18 | 22 | 11 | 14 | 13 | 90 |
| Expected | 15 | 15 | 15 | 15 | 15 | 15 | 90 |

Just looking at this data, we see that there appears to be too many orange and blue M&Ms in our sample and not quite enough of the others, but like flipping a coin 100 times and not getting exactly 50 heads, this may not be surprising. A chi-square distribution could help us examine if we are deviating far enough from what is expected given our observations to accept our alternate hypothesis and reject the null.

As in other tests of Type I error, we will also need to know the degrees of freedom (*df*), as the shape of a particular chi-square distribution is dependent on degrees of freedom. In chi-square analyses of single variables, as in our previous example, $df = k - 1$, where $k$ is the number of categories in the variable.

Now we can begin our analysis by stepping through our hypothesis test. We have already stated our hypotheses, so now it is time to set our α level. As is typical, we will set $\alpha \leq 0.05$.

Since our critical value will be based on *df*, we need to compute that. In this case, $df = 6 - 1 = 5$. Using Table 6 in Appendix F, we see that our critical value is 11.07. That is, any calculated $X^2 \geq 11.07$ will be considered statistically significant since this is the threshold at which our chance of making a Type I error is less than or equal to 5%. This is illustrated in Figure 9.1.

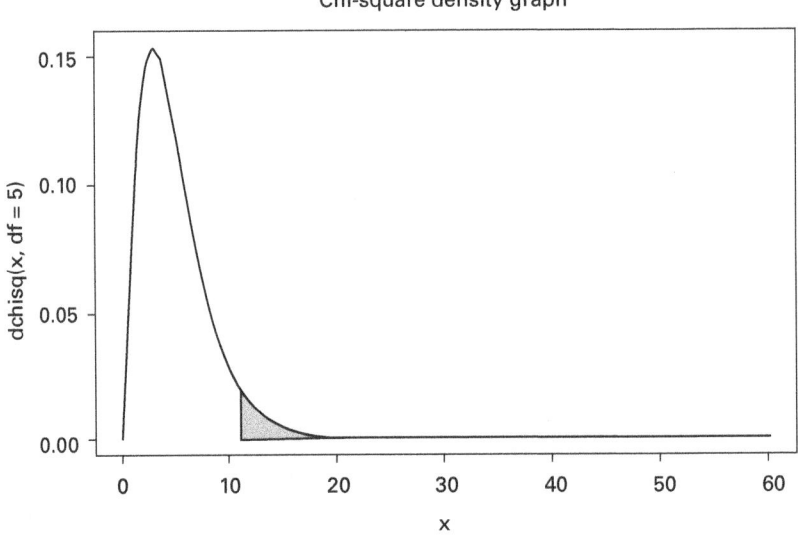

Chi-square density graph

FIGURE 9.1. Chi-square density graph with *df* = 5 and α ≤ 0.05.

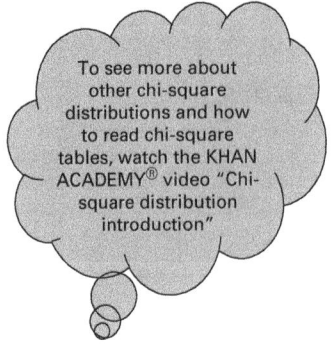

To see more about other chi-square distributions and how to read chi-square tables, watch the KHAN ACADEMY® video "Chi-square distribution introduction"

Notice that this distribution looks quite different than others we have seen previously, and as we move on throughout this chapter, you will see how *df* greatly impacts the overall shape of the $X^2$ distribution.

The formula for calculating Pearson's $X^2$ is shown in Formula 9.1:

$$\text{Formula 9.1:} \quad X^2 = \sum \frac{(f_o - f_e)^2}{f_e},$$

where $f_o$ is the frequency of what is observed and $f_e$ is the frequency of what is expected.

Now we can calculate our test statistic:

$$
\begin{aligned}
X^2 &= \frac{(12-15)^2}{15} + \frac{(18-15)^2}{15} + \frac{(22-15)^2}{15} + \frac{(11-15)^2}{15} + \frac{(14-15)^2}{15} + \frac{(13-15)^2}{15} \\
&= \frac{(-3)^2 + (3)^2 + (7)^2 + (-4)^2 + (-1)^2 + (-2)^2}{15} \\
&= \frac{9 + 9 + 49 + 16 + 1 + 4}{15} = \frac{88}{15} = 5.87
\end{aligned}
$$

Since our calculated $X^2$ is less than our critical value, we are unable to reject our null hypothesis.

Often, we are interested in the association between two variables. For example, we may want to know if smokers are more likely than nonsmokers to be admitted to the hospital through the emergency department:

> To see another example of a chi-square test, watch the KHAN ACADEMY® videos "Pearson's chi-square test (goodness of fit)," "Chi-square statistic for hypothesis testing," and "Chi-square goodness of fit example"

$H_0$: Smokers and nonsmokers are equally likely to be admitted to the hospital.
$H_1$: Smokers and nonsmokers are not equally likely to be admitted to the hospital.

As is typical, we will set $\alpha \leq 0.05$.

Here, we can begin to look for an association between these two categorical variables with the chi-square test by constructing a **contingency table**. As displayed in Table 9.3, we have information from the *hospital* dataset both on patients' smoking behavior and whether or not they were admitted through the emergency department. Table 9.3 was constructed in much the same way we created tables in Chapter 2. Here we entered the following into the Console:

```
> addmargins(table(hospital$edadmit, hospital$smoke))
```

We used the **addmargins()** function in the **table()** function because we wanted to get both column and row totals. For explanatory purposes, we labeled each of the cells in the interior of this table with the letters *a* through *d*.

TABLE 9.3. Table Showing Frequencies of Patients by Admission and Smoking

| | Smoker? | | |
|---|---|---|---|
| **Admit from Emergency Dept?** | **No** | **Yes** | **Totals** |
| **No** | 16 (*a*) | 8 (*b*) | 24 |
| **Yes** | 21 (*c*) | 5 (*d*) | 26 |
| **Total** | 37 | 13 | 50 |

To begin the analysis, a contingency table composed of two columns for *smoker* and two rows for *edadmit to* needs to be constructed. Because there are two columns and two rows, this table is referred to as a 2 × 2 contingency table. A table with two columns and three rows, on the other hand, would be a 2 × 3 contingency table. Our contingency table shows patients in four different categories:

- Smokers not admitted from the emergency department
- Smokers admitted from the emergency department
- Nonsmokers not admitted from the emergency department
- Nonsmokers admitted from the emergency department

To complete this problem and determine whether there is an association between these variables, we will need to calculate both a chi-square test statistic and *df*. We need to use a different formula for *df* than we did earlier since we are dealing with more than one variable. In that case, $df = (r – 1)(c – 1)$. In this case, then, $df = (2 – 1)(2 – 1) = 1$ and our critical value for chi-square is 3.84, which we looked up in Table 6. The shaded area in Figure 9.2 shows the chi-square distribution with 1 degree of freedom and the shaded area in which we would determine that our calculated chi-square was sufficiently unusual (i.e., values greater than 3.84) to determine them to be "statistically significant."

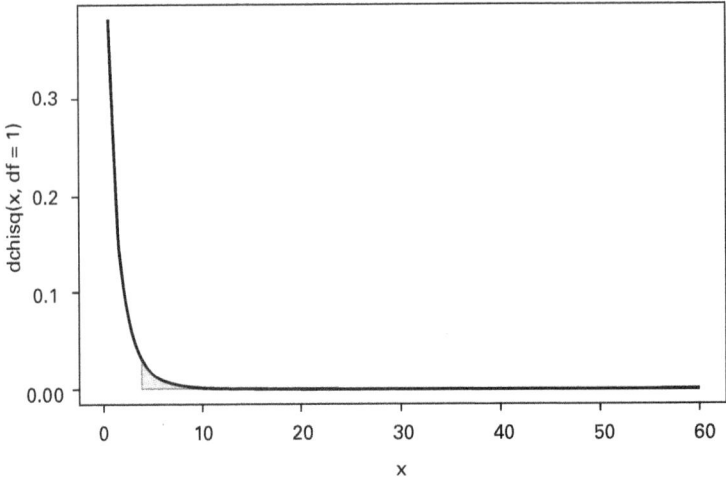

Chi-square density graph

FIGURE 9.2. Chi-square density graph with *df* = 1 and α ≤ 0.05.

Notice how the shape of this chi-square distribution looks compared to the distribution with 5 degrees of freedom in Figure 9.1.

Table 9.3 indicates that more people are admitted from the emergency department than not (26 compared to 24), and there are more nonsmokers admitted from this group than smokers. While you learned to calculate the chi-square using Formula 9.1, it is not necessarily intuitive how to determine what is an expected frequency in each cell when you have more than one variable. After all, while there is an equal number of smokers and nonsmokers, hospital admission is split unequally between admission from the emergency department and other sources. Therefore, when determining what an expected value is for each cell in a contingency table with two variables, we have to consider both variables, and we do so using Formula 9.2, shown again here:

$$\text{Formula 9.2}: \quad f_e = \frac{(row\ total)(column\ total)}{N}$$

Therefore, the expected frequency for cell $a$ would be:

$$f_e = \frac{(row\ total)(column\ total)}{N} = \frac{(24)(37)}{50} = 17.76$$

The expected frequency for cell $b$ would be:

$$f_e = \frac{(row\ total)(column\ total)}{N} = \frac{(24)(13)}{50} = 6.24$$

The expected frequency for cell $c$ would be:

$$f_e = \frac{(row\ total)(column\ total)}{N} = \frac{(26)(37)}{50} = 19.24$$

The expected frequency for cell $d$ would be:

$$f_e = \frac{(row\ total)(column\ total)}{N} = \frac{(26)(13)}{50} = 6.76$$

We can now calculate the chi-square for this contingency table using Formula 9.1 to compare the observed and expected frequencies for each cell:

$$X^2 = \sum \frac{(f_o - f_e)^2}{f_e} = \frac{(16-17.76)^2}{17.76} + \frac{(8-6.24)^2}{6.24} + \frac{(21-19.24)^2}{19.24} + \frac{(5-6.76)^2}{6.76}$$
$$= 0.17 + 0.50 + 0.16 + 0.46 = 1.29$$

Since our calculated $X^2$ of 1.29 is less than the critical value of 3.84, we are unable to reject the null hypothesis that there is no association between smoking and how people are admitted.

While this example illustrates calculating a chi-share test statistic in the simplest case, with a 2 × 2 table, the actual size of the table does not matter, and larger tables use the exact same procedure as illustrated previously.

### Yates's Correction

A rule of thumb is that no more than 25% of the expected cell values should be below 5. When this is not the case, it is necessary to make a correction. In terms of a 2 × 2 table, adding or subtracting 0.5 from expected values will correct for continuity. *R* automatically does this.

### FISHER'S EXACT TEST

For tables with small expected frequencies and 2 × 2 tables, Fisher's exact test should be used instead of the typical chi-square. It tests the probability that the observed frequencies would be exactly obtained under the null hypothesis in the population proportions. Whereas Pearson's chi-square is only an approximation of the observed frequencies that would be obtained under the null hypothesis in the population proportions, Fisher's exact test is a more conservative test and is not as impacted by small samples. The probability is displayed in Formula 9.3:

$$\text{Formula 9.3:} \quad p = \frac{(a+b)!(c+d)!(a+c)!(b+d)!}{N!a!b!c!d!}$$

As you can see, the computation of this test is quite complex and time consuming. Luckily Fisher's exact test only takes a few easy steps to run in *R*.

### USING *R* TO CALCULATE CHI-SQUARE

We can calculate a more accurate chi-square than our hand calculation in the previous example in *R* very easily by taking into account Yates's correction. To begin, we will need to sort the data in order to create the 2 × 2 contingency table that was developed in Table 9.3. We will use the **table()** function to store it in a vector that we will call *t1*.

```
> t1<-table(hospital$edadmit, hospital$smoke)
> t1
```

The output from this is shown in the Console:

```
        No Yes
  No   16   8
  Yes  21   5
```

*edadmit* is in the rows and *smoke* is in the columns.

To run the chi-square use the following *R* syntax, the output of which is shown in Figure 9.3:

```
chisq.test(t1)
```

Pearson's Chi-squared test with Yates'
continuity correction

data:  t1
X-squared = 0.66118, df = 1, p-value = 0.4161

FIGURE 9.3. Chi-square on 2 × 2 table with Yates's correction.

This output is easy to read with the output providing the $X^2$ test statistic and *df*. Notice that a critical $X^2$ value is not provided, but the probability of making a Type I error is. Therefore, before you can try to reject a null hypothesis, be clear at what level $\alpha$ is set.

You could just as easily do a Fisher's exact by entering the following into the Console:

```
> fisher.test(t1)
```

The results of this are shown as follows:

```
    Fisher's Exact Test for Count Data

data: t1
p-value = 0.3388
alternative hypothesis: true odds ratio is not
    equal to 1
95% confidence interval:
    0.103081 2.057050
sample estimates:
odds ratio:
    0.4834018
```

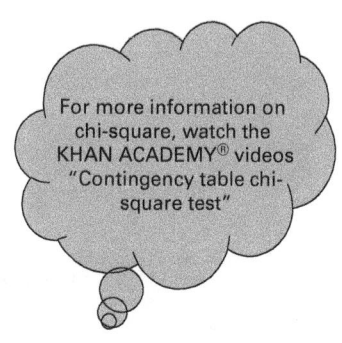

For more information on chi-square, watch the KHAN ACADEMY® videos "Contingency table chi-square test"

Here, the calculated probability of making a Type I error is about 34%, but now we have corrected for small expected cell sizes, if that was a concern.

**Another Way to Do $X^2$ With *R*: Introducing *gmodels***

The *gmodels* package utilizes a function, `CrossTable()`, that will allow us to get all of the information we want to get for a $X^2$ test in one command. Be sure to load this package before you begin, and we will do a slightly more complicated example than the one earlier.

Continuing to look at hospital patients, let's consider whether patients who live with a spouse or partner may have different discharge plans than those who do not:

$H_0$: Living with a spouse or partner is not related to discharge plan.
$H_1$: Living with a spouse or partner is related to discharge plan.

To test this hypothesis using *gmodels*, enter the following into the Console to view the output displayed in Figure 9.4:

```
> CrossTable(hospital$discharge, hospital$cohabit,
prop.r=F, prop.c=F, prop.t=F, prop.chisq=F,
expected=T, chisq=T)
```

```
   Cell Contents
|-----------------------|
|                     N |
|            Expected N |
|-----------------------|


Total Observations in Table:  50


                 | hospital$cohabit
hospital$discharge |        No |       Yes | Row Total |
-------------------|-----------|-----------|-----------|
            Home |        10 |        17 |        27 |
                 |    12.420 |    14.580 |           |
-------------------|-----------|-----------|-----------|
         Nursing |        10 |         3 |        13 |
                 |     5.980 |     7.020 |           |
-------------------|-----------|-----------|-----------|
           Rehab |         3 |         7 |        10 |
                 |     4.600 |     5.400 |           |
-------------------|-----------|-----------|-----------|
    Column Total |        23 |        27 |        50 |
-------------------|-----------|-----------|-----------|


Statistics for All Table Factors


Pearson's Chi-squared test
------------------------------------------------------------
Chi^2 =  6.908258      d.f. =  2      p =  0.03161482
```

FIGURE 9.4. Chi-square test using *gmodels*.

The **CrossTable()** function itself can provide a lot of output. Here, as illustrated at the top of Figure 9.4, we have limited what is displayed in each cell to display the frequency of each followed by the expected frequency. Therefore, for example, we observed 10 people who didn't cohabit with anyone and were discharged home, but 12.42 people were expected. To show expected frequencies, we used the **expected=T** option. Similarly, since we wanted to show the chi-square hypothesis test, we used the **chisq=T** option. Since we only wanted these options displayed, we suppressed the other default options by marking them as **F** for False.

Notice that our *df* has increased as we now have a 2 × 3 table. Our calculated test statistic is 6.91 and the *p*-value is 0.03161482. Thus, if α is set at 0.05, we can reject the null hypothesis and accept the alternate.

There appears to be a relationship between living with a spouse or partner and discharge plan, but what is that relationship? We need to look at our contingency table further. Notice that the vast majority of those who cohabit with a spouse or partner are discharged home, and only a few are sent to a nursing home. However, when you look at people who do not live with a spouse or partner, those folks seem more likely to be discharged to a nursing home. It might be helpful to look at this a little more closely, perhaps by getting percentages by column; however, before we do that, we should also notice that there are small expected cell sizes for those going to rehab centers. In the event of small cell sizes, we will also want to look at Fisher's exact, which corrects for these small expected values. To do this, then, we can rerun the **CrossTable()** function but change some of our options to get those column proportions and Fisher's exact:

```
> CrossTable(hospital$discharge, hospital$cohabit,
prop.r=F, prop.c=T, prop.t=F, prop.chisq=F, chisq=T,
fisher=T)
```

Here, we added the options **fisher=T** and **prop.c=T** to get the output shown in Figure 9.5. We also dropped the option to show expected values.

Again, notice the key to the cell contents in the top of the figure. With column proportions added, we can see from the column totals that 54% of the sample lived with a spouse or partner compared to 46% who did not. We also see that of all the people who lived with a spouse or partner, 63% were discharged home, 25.9% were sent to a rehab center, and 11.1% were discharged to a nursing home. This is different than those who didn't live with a spouse or partner. Many more of them (43.5%) were discharged to nursing homes, and less were sent to rehabilitation centers. It seems that those who don't live with a significant other are more likely to go to nursing facilities, while those who do live with a significant other are most likely to be discharged home.

```
   Cell Contents
|-------------------------|
|                     N  |
|         N / Col Total  |
|-------------------------|


Total Observations in Table:  50


                    | hospital$cohabit
hospital$discharge  |      No  |      Yes  | Row Total  |
--------------------|----------|-----------|------------|
              Home  |     10   |     17    |     27     |
                    |  0.435   |   0.630   |            |
--------------------|----------|-----------|------------|
           Nursing  |     10   |      3    |     13     |
                    |  0.435   |   0.111   |            |
--------------------|----------|-----------|------------|
             Rehab  |      3   |      7    |     10     |
                    |  0.130   |   0.259   |            |
--------------------|----------|-----------|------------|
      Column Total  |     23   |     27    |     50     |
                    |  0.460   |   0.540   |            |
--------------------|----------|-----------|------------|


Statistics for All Table Factors


Pearson's Chi-squared test
------------------------------------------------------------
Chi^2 =  6.908258     d.f. =  2     p =  0.03161482


Fisher's Exact Test for Count Data
------------------------------------------------------------
Alternative hypothesis: two.sided
p =  0.04204025
```

FIGURE 9.5. CrossTable chi-square with Fisher's exact and column proportions.

## MEASURING THE STRENGTH OF THE RELATIONSHIP

So far, we have been concerned only with testing the existence of a relationship between two categorical variables. We set up and tested whether the null hypothesis can
be rejected. We also established the degree of risk of Type I error we were willing to
accept and the differences we observed in our findings. In this section, you will be
introduced to a number of statistics used in contingency tables to quantify the degree
of the strength of the relationship between two categorical variables.

It would seem logical that you could determine the strength of the relationship between two variables by the significance level of a relationship. For example, if we compare a significance level of 0.025 to 0.05, based on this idea, the former would be considered a stronger relationship.

Keep in mind, however, that when sample sizes are large, even a trivial relationship will be statistically significant. For example, compare Tables 9.4 and 9.5.

TABLE 9.4.  $n = 200$

| | |
|---|---|
| 53 (53%) | 48 (48%) |
| 47 (47%) | 52 (52%) |

$X^2 = 0.50$; $p = 0.48$.

TABLE 9.5.  $n = 20,000$

| | |
|---|---|
| 5300 (53%) | 4800 (48%) |
| 4700 (47%) | 5200 (52%) |

$X^2 = 50$; $p = 0.00$.

Notice that Table 9.5 has exactly 100 times more observations in each cell than Table 9.4, so the tables have the same exact differences in the column proportions (53% compared to 47%, and 48% compared to 52%). The chi-square is nonsignificant in Table 9.4; nevertheless, it is significant with the larger sample size in Table 9.5. Also notice that the calculated test statistic is proportional to the increase in the sample size. Even though Table 9.5 has 100 times more observations than Table 9.4, the same degree of difference can be found in both tables, with the chi-square being 100 times greater in Table 9.5.

Given the issues described, perhaps a more essential issue is, if a statistically significant relationship exists, how strong is it?

## CHI-SQUARE-BASED MEASURE OF THE STRENGTH OF ASSOCIATION

A widely used measure of association that can only be used with 2 × 2 tables is *phi* (ф), which considers that chi-square is proportional to the sample size. The formula for *phi* is shown in Formula 9.4:

$$\text{Formula 9.4}: \quad \phi = \sqrt{\frac{X^2}{n}}$$

As an example, let us examine $\phi$ in the example shown in Tables 9.4 and 9.5:

For Table 9.4 $\phi = \sqrt{\dfrac{5}{200}} = \sqrt{0.0025} = 0.05$

For Table 9.5 $\phi = \sqrt{\dfrac{50}{20000}} = \sqrt{0.0025} = 0.05$

You will notice that the value for $\phi$ is identical for both tables. Although the chi-square in Table 9.5 is 100 times as large as Table 9.4, the degree of association as measured by $\phi$ is the same. When there is absolutely no relationship between two variables, $\phi$ will equal zero. If the two variables have a perfect relationship, $\phi$ will equal 1. Given this, the relationships between the variables in both tables are very weak.

As another example, we can now look at how strong the relationship is between whether patients in the *hospital* dataset cohabit and their discharge plan, as illustrated in Figures 9.4 and 9.5:

$$\phi = \sqrt{\dfrac{X^2}{n}} = \sqrt{\dfrac{6.908}{50}} = \sqrt{0.138} = 0.372$$

The relationship here is quite a bit stronger one than what we observed in Tables 9.4 and 9.5.

Another measure of association is Cramer's V. The formula is presented in Formula 9.5:

$$\text{Formula 9.5:} \quad Cramer's\ V = \sqrt{\dfrac{x^2}{n(k-1)}}$$

In this formula $n$ = the total number of observations and $k$ is the lesser of the number of rows or columns in a table. For a $2 \times 2$ table, Cramer's V and $\phi$ will be identical.

To look at the association between cohabitation and discharge plan, as illustrated earlier, with Cramer's V, we can do the calculation that follows:

$$Cramer's\ V = \sqrt{\dfrac{x^2}{n(k-1)}} = \sqrt{\dfrac{6.908}{50(2-1)}} = \sqrt{\dfrac{6.908}{50}} = \sqrt{0.138} = 0.372$$

In this example, *Cramer's V* and $\phi$ were identical because *cohabit* had the fewest number of categories at two. The interpretation of *Cramer's V* is identical to that of $\phi$ with possible values ranging from 0, indicating total independence, to 1, indicating total dependence.

Yet another measure of association based on chi-square is the contingency coefficient. The formula is presented in Formula 9.6. Once again, using the example of the relationship between cohabitation and discharge plan, we provide an example of how it is calculated:

$$\text{Formula 9.6}: \quad C = \sqrt{\frac{x^2}{n + x^2}} = \sqrt{\frac{6.908}{50 + 6.908}} = \sqrt{\frac{6.908}{56.908}} = \sqrt{0.121} = 0.348$$

Like Cramer's $V$ and phi, $C$ is 0 when the variables are independent. The upper limit of $C$ will always be less than 1. For this reason, the contingency coefficient can be more difficult to interpret because the upper limit of $C$ increases as the size of the table increases.

## USING THE *R VCD* PACKAGE TO CALCULATE MEASURES OF STRENGTH OF ASSOCIATION

It is easy to calculate measures of association using the *vcd* package in *R*. To illustrate this, let us go into more depth on patients' relationship status and their discharge plans. We may not simply wonder about whether a patient is living with a spouse or partner, but his or her actual marital status.

The null hypothesis is: There is no relationship between patients' marital status and where they are discharged. The alternative hypothesis is: There is a relationship between patients' marital status and where they are discharged.

As illustrated earlier, we could compute a $X^2$ in *R* by first building a table. We entered the following into the Console and the table itself is displayed in Figure 9.6:

```
> t2<-table(hospital$discharge, hospital$marital)
> t2
```

```
        Divorced Married Single Widowed
Home           5      17      5       0
Nursing        2       2      5       4
Rehab          4       4      2       0
```

FIGURE 9.6. Two-way table of marital status and discharge location.

We can now obtain the $X^2$ by entering the following in the Console, with the result being illustrated in Figure 9.7:

```
> chisq.test(t2)
```

```
            Pearson's Chi-squared test

data:  t2
X-squared = 19.218, df = 6, p-value = 0.00381
```

FIGURE 9.7. Chi-square of relationship between marital status and discharge location.

We see that the relationship is significant with $p = 0.0038$, but how strong is the relationship?

The *vcd* package's `assocstats()` function can be employed to calculate chi-square and the measures of association described previously. After loading the *vcd* package, you can use the vector we created to calculate the chi-square to look at our association coefficients, with the result being illustrated in Figure 9.8.

```
> assocstats(t2)

                          X^2 df  P(> X^2)
    Likelihood Ratio 18.973  6 0.0042088
    Pearson          19.218  6 0.0038104

    Phi-Coefficient    : NA
    Contingency Coeff.: 0.527
    Cramer's V         : 0.438
```

FIGURE 9.8. Results of *vcd* package `assocstats()` function.

The column labeled $X^2$ in Figure 9.8 displays the chi-square values, and that labeled $P(> X^2)$ shows the significance values for chi-square. The chi-square next to the row with "Pearson" displays what you have learned to calculate and is the same as what is illustrated in Figure 9.7. Because the contingency table was larger than 2 × 2, a phi-coefficient couldn't be calculated. The other two measures of association are similar and show a moderately strong relationship between marital status and discharge plan.

## LAMBDA

Lambda ($\lambda$) belongs to a class of measures of association known as proportional reduction of error (PRE), and it is frequently used when both variables are nominal. The idea is to better understand how much the addition of an independent variable does in reducing the error in the prediction of the dependent variable. PRE measures are considered superior because they are not based on chi-square. This is because they are impacted by sample size. The formula is presented in Formula 9.7:

$$\text{Formula 9.7:} \quad \lambda = \frac{E1 - E2}{E1}$$

$E1$ represents the errors in predicting the dependent variable without including the independent variable. It is the total sample size minus the largest row total. $E2$ is the sum of the difference of each column's largest cell less the column's total $n$. For illustrative purposes, we can calculate $\lambda$ for the data illustrated in Figure 9.6. To simplify, we have created Table 9.6 with both column and row totals so we can do our calculation.

TABLE 9.6. Table With Patients' Marital Status and Discharge Plans

| Discharge | Marital Status | | | | |
|---|---|---|---|---|---|
| | Divorced | Married | Single | Widowed | Total |
| Home | 5 | 17 | 5 | 0 | 27 |
| Nursing | 2 | 2 | 5 | 4 | 13 |
| Rehab | 4 | 4 | 2 | 0 | 10 |
| Total | 11 | 23 | 12 | 4 | 50 |

$E1 = n -$ largest row total $= 50 - 27 = 23$

$E2 =$ for each column, subtract the largest cell frequency from the column total $= (11 - 5) + (23 - 17) + (12 - 5) + (4 - 4) = 6 + 6 + 7 + 0 = 19$

$$\lambda = \frac{E1 - E2}{E1} = \frac{23 - 19}{23} = \frac{4}{23} = 0.174$$

The addition of the independent variable *marital* reduced the error in the prediction of *discharge* by 17.4%.

## USING *R* TO CALCULATE LAMBDA

Lambda can be calculated in *R* with the `lambda.test()` function in the *rapportools* package. Once that package is loaded, we can re-create the calculation we did earlier. Remember that we previously created the vector *t2* that held a table with the variables *marital* and *discharge*. We will use that vector in our function. Enter the following into the Console:

```
> lambda.test(t2)
```

The results are produced in Figure 9.9.

```
$row
[1] 0.1111111

$col
[1] 0.173913
```

FIGURE 9.9. Example of calculating lambda with *rapportools*.

The lambda for $col is the value we are interested in because the independent variable, *marital*, is displayed in the columns. The results are exactly as we hand-calculated previously. This is congruent with what was found in the chi-square measures of association; there is a moderate association between the independent and dependent variable. Here, however, we can add a specific reduction in error in predicting discharge plan from marital status.

## GAMMA

Often it is necessary to calculate the association between two variables that are measured at the ordinal level. Gamma ($\gamma$) is the preferred method of measuring the strength and direction of an ordinal variable arranged in a contingency table or with dichotomous nominal variables. It is also a PRE measure of association.

To illustrate this, we will use the *hospital* dataset to consider whether patients' satisfaction with their nurses while in the hospital (*nurse*) is related to their satisfaction with their overall experience (*overall*) while in the hospital.

To begin the analysis, we will construct a contingency table with *nurse* in the columns and *overall* in the rows. This table will be a 2 × 2 table with satisfaction with nursing and overall satisfaction having been dichotomized as either high (H) or low (L). You can construct a table like the one shown in Table 9.7 simply by counting the number of people in each of four categories:

- High satisfaction with nursing and high overall satisfaction
- High satisfaction with nursing and low overall satisfaction
- Low satisfaction with nursing and high overall satisfaction
- Low satisfaction with nursing and low overall satisfaction

TABLE 9.7. Categories of Satisfaction With Nursing by Overall Satisfaction

| Satisfaction with Nurse | Overall Satisfaction | | Totals |
|---|---|---|---|
| | High | Low | |
| High | 19 | 9 | 28 |
| Low | 8 | 14 | 22 |
| Total | 27 | 23 | 50 |

With $\gamma$, we try to predict the order of *pairs of cases*. That is, we want to predict whether one case will have a higher or lower score than another. For example, if a patient scores low on nursing satisfaction and low on overall satisfaction, will the next patient also score low on both variables?

The calculation for gamma is presented in Formula 9.8:

$$\text{Formula 9.8}: \quad \gamma = \frac{n_c - n_d}{n_c + n_d}$$

First, we must calculate $n_c$, which represents the sum of all concordant pairs. To calculate this in this $2 \times 2$ table, start with the *High-High* cell at the upper left and multiply the cell frequency by the sum of all the cell frequencies below and to the right of it. In this case, there is only one cell that is below and to the right of the High-High cell:

$$n_c = 19(14) = 266$$

Then, we will calculate $n_d$, which represents the sum of all discordant pairs. In this $2 \times 2$ table, we will start with the High-Low combination, and then multiply that cell frequency by the sum of all the cell frequencies above and to the right of it. Again, there is only one cell here that meets this criterion due to the table size:

$$n_d = 9(8) = 72$$

Now we can calculate gamma as follows:

$$\gamma = \frac{C-D}{C+D} = \frac{266-72}{266+72} = \frac{194}{338} = 0.574$$

A gamma between 0.0 and 0.30 is considered to indicate a weak association, 0.31 to 0.60 denotes a moderate relationship, and 0.61 or higher is indicative of a strong relationship. Knowing a patient's satisfaction with nursing increases our predication of their overall satisfaction by 57.4%.

## USING *R* TO CALCULATE GAMMA

We can use *R* to calculate gamma easily, which is particularly helpful because tables that are larger than $2 \times 2$ can become labor intensive. We will re-create the hand calculation we completed earlier by using the **GKgamma()** function in the *vcdExtra* package.

To begin, we need to create a table for both the chi-square test and the gamma function. Enter the following in the Console:

```
> t3<-table(hospital$nurse, hospital$overall)
```

Next, we will use the **chisq.test()** to test if a relationship exists using the following syntax:

```
>chisq.test(t3)
```

The output displayed in Figure 9.10 should appear in the Console.

```
            Pearson's Chi-squared test with Yates' continuity correction

data:  t3
X-squared = 3.7331, df = 1, p-value = 0.05334
```

FIGURE 9.10. Chi-square test between satisfaction with nursing care and overall hospital satisfaction.

The results indicate that a statistically nonsignificant ($p > 0.05$) relationship exists between *nurse* and *overall*. Because both variables are dichotomous, gamma can be used to test the strength of the relationship between these variables.

With the *vcdExtra* package loaded, use the following syntax:

```
> GKgamma(t3)
```

The output illustrated in Figure 9.11 is produced in the Console.

```
> GKgamma(t3)
gamma       : 0.574
std. error  : 0.201
CI          : 0.18 0.968
```

FIGURE 9.11. Gamma for satisfaction with nursing and overall satisfaction.

Notice that the output for *gamma* is exactly as we calculated previously.

## ANOTHER HYPOTHESIS TEST: MCNEMAR'S TEST

McNemar's test assesses if a statistically significant change in proportions has occurred on a dichotomous variable measured at two time points with the same subjects or any matched pairs of observations, such as two people observing the same event. Therefore, it can only be applied to a $2 \times 2$ contingency table.

As an example, suppose a high school administrator wants to determine if an intervention has an effect on whether or not at-risk students intend to obtain a high school diploma. At the start of the study, guidance counselors ask at-risk students whether they plan to quit school before they graduate, and after the intervention, they ask them the same question. The null hypothesis assumes that there is no change in intention to quit after the intervention, and the alternative hypothesis assumes that students' intentions to quit change after the intervention.

A count of the students' responses is recorded, and a $2 \times 2$ table is created, as illustrated in Table 9.8. Notice that we have labeled the cells *a* through *d* for explanatory purposes.

McNemar's test is conducted using the chi-square test statistic to determine whether the intervention has a statistically significant association with the number of students considering quitting school.

TABLE 9.8. Quitting School: Thinking of Leaving Before and After Intervention

| Plan on Quitting HS: Before Intervention | Plan on Quitting HS: After Intervention | | |
|---|---|---|---|
| | No | Yes | Totals |
| No | 8(a) | 4(b) | 12 |
| | 40% | 40% | |
| Yes | 12(c) | 6(d) | 18 |
| | 60% | 60% | |
| Total | 20 | 10 | 30 |

As Table 9.8 illustrates, of the 18 students who initially planned to quit school, 12 (60%) now intend on remaining until graduation.

The formula for McNemar's test is presented in Formula 9.9:

$$Formula\ 9.9:\ X^2 = \frac{(b-c)^2}{b+c}$$

Since we are working with a 2 × 2 table, we have $df = 1$. We can now calculate the chi-square as follows:

$$X^2 = \frac{(b-c)^2}{b+c} = \frac{(4-12)^2}{4+12} = \frac{(-8)^2}{16} = \frac{64}{16} = 4.0$$

Using the Table 6 in Appendix F, with 1 degree of freedom, a chi-square of 3.84 or higher is needed with α set at 0.05. The calculated chi-square is greater than that critical value, so the null hypothesis can be rejected and the alternate accepted.

## USING *R* TO CONDUCT MCNEMAR'S TEST

We can re-create what we did previously in *R*. To begin this analysis, we will use the **matrix()** function to enter the frequencies of the cells in the order of *a, b, c,* and *d*. The number of columns are entered using the **ncol=2** option. The **byrow=T** is added to reflect the order of the cells in the table. The **correct=F** is added to replicate the hand calculation in which a continuity correction was not added. The output of this is shown in Figure 9.12.

```
>quit <- matrix(c(8, 4, 12, 6), ncol=2, byrow=T)
>mcnemar.test(quit,correct=F)
```

```
McNemar's Chi-squared test

data:  quit
McNemar's chi-squared = 4, df = 1, p-value = 0.0455
```

FIGURE 9.12  Using R to calculate McNemar chi-square in a 2 × 2 table.

Rerunning this function excluding the **correct=F** option yields a nonsignificant McNemar chi-square as shown in Figure 9.13.

```
McNemar's Chi-squared test with continuity correction

data:  quit
McNemar's chi-squared = 3.0625, df = 1, p-value = 0.08012
```

FIGURE 9.13  McNemar chi-square with correction.

Of course, if you had previously created a table and stored it in a vector, you could use that in a **mcnemar.test()** function.

## CONCLUSION

In this chapter, we looked at how you could examine associations when variables are categorical. Tests of Type I error use the chi-square distribution and test statistic to determine if observations are significantly different from what is expected. Contingency tables are used to help you sort data into categories based on how each variable is operationalized and observed. If it is unlikely that a Type I error has been made, you can assess how strong of a relationship is detected between your variables, keeping in mind that just because a relationship is statistically significant does not mean that the relationship between the variables is strong. Additional measures such as phi, C, or lambda can help you ascertain the strength of the relationship between variables. In all cases, and as we have discussed throughout this book, it is important to choose measures and statistical tests that are appropriate given the nature of your data. You will want to review, therefore, when it is appropriate to use any given analytical technique from time to time or if you are unsure which is most appropriate.

## EXERCISES

1. You and your friends are not happy with the food served in the dining hall. It seems like several of the menu items are not touched, and there is never enough of your favorite meal, sushi. After meeting with the director of dining

services, you decide to take a random sample of 120 students and ask them what their favorite menu item is. The responses you get are as follows:

| Menu Item | Frequency of Preference |
|---|---|
| Sushi | 36 |
| Chef's salad | 28 |
| Pizza | 30 |
| Burger | 26 |

   a. Write both null and alternate hypotheses to test whether there is, in fact, a preference for one menu item over others.

   b. Conduct a test of Type I error at the 0.05 level to test your hypothesis. Be sure to specify the critical test statistic and degrees of freedom.

   c. Can you make a case to the director of dining services to serve more sushi? Why or why not?

For the next exercises, use the *student* data provided with this book:

As part of your job in the Provost's Office, you are asked to begin building a profile of the students in each college. Because you have already collected data from 20 students, you decide to use this information to start building your profile. For each of the following questions, determine if there is a significant relationship between an independent variable and your dependent variable, *College*. Because you have a small sample size, you set your alpha level at 0.10. If a significant relationship exists, test the strength of that relationship with an appropriate measure and explain your findings.

  2. Do first-generation students have a preference for college?

  3. Does the dormitory in which students live impact the college in which students take classes?

  4. Is students' gender related to college preference?

  5. Students at your university have the option to attend a series of workshops to help them improve their memory, with the idea that improved memory will increase grades. Did the workshops significantly improve students' memories?

## *R* EXERCISES

For the following exercises, use the *hospital* dataset provided with this book.

You have been tasked with identifying factors related to patients being readmitted to the hospital within 30 days of discharge. Because your dataset is relatively small, you have been asked to test each independent variable at both the 0.05 and 0.10 levels. If a statistically significant relationship exists, you should provide a measure of how strong the relationship is.

1. Is the gender of patients related to readmission within 30 days?
2. Is smoking related to readmission?
3. Is primary care position related to readmission?
4. Is location of discharge related to readmission?

# /// 10 /// POWER ANALYSIS

To work through the examples in this chapter, you will need to *install* and *load* the following *R* packages:

- *pwr*

For more information on how to do this, refer to Appendix C.

## INTRODUCTION

In this chapter, you will be introduced to the concepts of power analysis. Calculating power analysis by hand is extremely time consuming, so the focus here will be on utilizing *R* to conduct the analysis with the *pwr* package. The package is based on Jacob Cohen's (1988) work on power analysis.

**Power analysis** deals with the probability of correctly detecting a difference or relationship when one exists. Another way of stating this is that statistical power is the probability of correctly rejecting the null hypothesis.

As an example, let's consider testing the effect of an intervention to reduce anxiety in two groups. In the experiment, one group receives the intervention to reduce anxiety, and the other group does not. The power would be the probability of being able to detect a difference between the two groups if one existed. If there were a power of 0.9, this would tell us that, given the same sample size, a statistically significant difference between groups would be detected 90% of the time if one actually existed.

Power is directly related to sample size, where larger samples have more power to detect a statistical difference. Additionally, the strength of the relationship can also impact sample size. The stronger the relationship is, the smaller the sample size needed to detect a difference.

There are situations in which setting a lower power level is desired. This could occur, for example, if there are a limited number of subjects available or your budget to conduct research is restricted. Acceptable levels of power vary between 0.8 and 0.9. Later in the chapter, we show how you can graphically determine how sample size would vary with higher and lower power.

## WHY CONDUCT A POWER ANALYSIS?

For Sal Khan's explanation on the basics of power, watch the KHAN ACADEMY® video "Introduction to power in significance tests"

There are several purposes for utilizing power analysis. As described in the previous example above, the most common use of this technique is to determine the number of subjects needed to identify an effect of a given size. Power analysis can also be employed to determine statistical power given an effect size and the number of participants available. For instance, if a researcher knows that there are only 50 subjects available to study an intervention, conducting a power analysis would be helpful in deciding if the study was worth doing. Studies with higher power are more likely to detect effects, while those with lower power are more likely to lead researchers to dismiss a potentially important effect (Murphy, Myors, & Wolach, 2014).

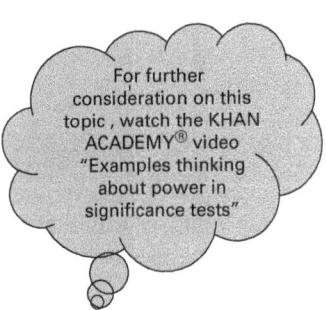

For further consideration on this topic, watch the KHAN ACADEMY® video "Examples thinking about power in significance tests"

Many grant funders require the inclusion of a power analysis in research proposals. Very often, it is cost effective to use a smaller sample, if possible. Another reason for doing so is ethical; why should participants be subjected to a research protocol if it is not necessary? Finally, being able to estimate an effect size with actual data or from the literature indicates familiarity with the research focus and allows grant funders to know of your expertise in the field.

## GETTING STARTED WITH THE *PWR* PACKAGE

There are four parameters included in a power analysis. At least three of the parameters need to be known to calculate the other:

- Sample size (*n*)
- Significance level (0.01, 0.05, etc.)
- Desired power (0.8, 0.9, etc.)
- Effect size (differences between groups)

As in the example presented earlier, if a researcher wants to know the number of subjects needed to correctly detect an effect 90% of the time (power = 0.9), the sample size parameter is left out of the calculation, and sample size will be estimated. On the other hand, if a researcher knows the number of subjects available to conduct a study, then the power parameter may be left out and it can then be calculated.

The researcher also needs to consider that different types of effect sizes are necessary under different research situations. In the example presented previously, the effect size would be based on differences in means between two groups. We will cover calculating power for the following types of effect sizes: *t*-tests, one-way analysis of variance, proportions, chi-square, and correlations. As a result, the effect size used is based on the type of statistical test you plan for in your study. This also implies that you will need to have some sort of information to estimate effect size. Later in this chapter we will discuss how to deal with situations where you do not have enough information to calculate effect size.

## Calculating Power for the *t*-Test

There are a number of functions available in the *pwr* package. The basic command syntax is as follows: `pwr.t.test(n =, d = sig.level=, power = alternative, type =)`. The formula for effect size, *d*, was presented earlier as Formula 7.11:

$$d = \frac{(\bar{X}_1 - \bar{X}_2)}{s_{pooled}}$$

The other parameters in this function are described as follows:

- *n* is the number of cases.
- *sig.level* is the significance level.
- *power* is the desired power level to be tested.
- *type* selects if the *t*-test is two groups (*"two.sample"*), a one-sample test (*"one.sample"*), or a dependent *t*-test (*"paired"*).
- *alternative* indicates if the *t*-test is nondirectional (*"two.sided"*) or directional (*"less"* or *"greater"*)

The required sample size is influenced by the magnitude of the effect, alpha, and whether you test a directional or nondirectional hypothesis. The larger the desired, anticipated, or observed effect is that needs to be detected, the smaller the sample size required to correctly reject the null and accept the alternative hypothesis. Conversely, the smaller the effect is that needs to be detected, the larger the sample size.

Let's consider the example discussed in Chapter 7 comparing mean differences in activities of daily living (ADL) scores between patients who receive one of two interventions, Treatment A or Treatment B. We want to know if we have enough research participants in each group to obtain a power of 0.9 with a significance level of 0.05 and a test of a nondirectional hypothesis. For the purpose of this example, we can use the effect size we calculated in Chapter 7, $d = 3.215$.

Load the *pwr* package and type the following into the Console:

```
pwr.t.test(d=3.215, sig.level=0.05, type="two.
sample", alternative="two.sided", power = 0.9)
```

The results are displayed in Figure 10.1.

```
        Two-sample t test power calculation

                 n = 3.353676
                 d = 3.215
         sig.level = 0.05
             power = 0.9
       alternative = two.sided

  NOTE: n is number in *each* group
```

FIGURE 10.1. Two-sample *t*-test power calculation.

With an effect size of 3.215 and a significance level of 0.05, an *n* of 4 in each group would be needed to obtain a power of 0.9 (i.e., the probability of detecting a statistically significant difference when it exists 90% of the time). Notice that we need to round up to the next highest integer since there is no such thing as 3.35 of a person. Sample sizes obtained with a power analysis designed to detect a medium or large effect size (as in this example, a large effect size) are inadequate in the sense that the precision of parameter estimates is insufficient. As in this example, an *n* of 4 is too small to compute accurate confidence intervals.

As mentioned, power analysis can also be employed to determine statistical power given an effect size and the number of available research subjects. Let's consider conducting a study where you have 50 subjects available. To go forward with the study, you determine that it is necessary to have the power of detecting a statistically significant difference when it exists 80% of the time. Based on the literature, you estimate that the effect size is 0.63. To follow is what you would enter into the Console. Notice that the power level is left out and the *n* of 50 is included:

```
>pwr.t.test(n=50,.63, sig.level=.05, type="two.
sample",alternative="two.sided")
```

The results shown in the Console are presented as follows:

```
Two-sample t test power calculation
              n = 50
              d = 0.63
        sig.level = 0.05
          power = 0.8767763
    alternative = two.sided
  NOTE: n is number in *each* group
```

In this case, the probability is 88% for correctly rejecting the null hypothesis based on the criteria established previously. This finding indicates, based on the number of subjects available, that there is adequate statistical power to conduct the study.

## Calculating Power for One-Way Analysis of Variance

Power for a one-way ANOVA is calculated with the `pwr.anova.test()` function. The general method for entering this in *R* is shown as follows:

```
>pwr.anova.test(k=, n=, f=, sig.level=, power=)
```

Here, *k* is the number of groups in the independent variable and *f* is the effect size. The calculation for effect size for ANOVA is shown as Formula 10.1:

$$\text{Formula 10.1:} \quad f = \sqrt{\frac{\sum p_i(\bar{X}_i - \bar{X})^2}{s^2}} \quad \text{and} \quad p_i = n_i / N$$

Cohen (1988) suggested the following guidelines for assessing the *f*: 0.01 is a small effect, 0.3 is a medium effect, and 0.5 is a large effect.

Here is how you would estimate the number of observations needed for a study that has four groups, an effect size of 0.37, a significance level of 0.05, and a power of 0.8. In this example, we chose $f = 0.37$ because we would like to be able to detect a moderate effect size. To follow is the command syntax you would use:

```
>pwr.anova.test(k=4, f=.37, sig.level=.05, power=.8)
```

The output produced by the command is shown in Figure 10.2.

```
   Balanced one-way analysis of variance power calculation

               k = 4
               n = 20.91279
               f = 0.37
       sig.level = 0.05
           power = 0.8

      NOTE: n is number in each group
```

FIGURE 10.2. Power analysis example for one-way ANOVA.

The estimated sample size is 21 observations per group for a total of 84 subjects, and the probability of correctly rejecting the null hypothesis is 80%.

### Calculating Power for Proportions

In some instances, we are interested in studies in which we have a dichotomous definition of success in much the same way as the examples of rolling dice or flipping coins illustrated in Chapter 3. In these cases, we want to know how the proportion of success changes over time. For example, we may want to know if the proportion of students being admitted to their first-choice law school changes when students are offered a new LSAT prep class.

Power for proportions is calculated with the **pwr.2p.test()** function. The general method for entering this in *R* is:

```
>pwr.2p.test(h=, n=, sig.level=, power=)
```

In this formula, *h* is the effect size and *n* is the shared sample size for each proportion, $P_1$ and $P_2$. The calculation for this effect size is displayed in Formula 10.2:

$$\text{Formula 10.2}: \quad h = 2\arcsin(\sqrt{P_1}) - 2\arcsin(\sqrt{P_2})$$

Effect size (*h*) can be calculated with the function **ES.h()**, which is also part of the *pwr* package. The **alternative()** option can also be included to specify a two-tailed ("*two-sided*") or one-tailed ("*less*" or "*greater*") test.

Let's consider the example of an intervention designed to increase the number of prelaw students who are ultimately admitted to their first-choice school by introducing a new LSAT prep class. We will need a sample of students to take the new course and another sample to take the existing course. The goal is to increase first-choice admission by 15%, from 43%, which is what it is now, to 58%. To run our study well, we want to know how many subjects we need to recruit in each group (i.e., an intervention group and a comparison group) with $\alpha = 0.05$ and a power of 0.9 (i.e., the probability of correctly rejecting the null 90% of the time). The following syntax will estimate the number of participants needed to test this intervention, and the output is displayed in Figure 10.3:

```
>pwr.2p.test(h=ES.h (.43, .58) , sig.level=.05,
power =.9)
```

```
> pwr.2p.test(h=ES.h (.43, .58) , sig.level=.05, power =.9)

     Difference of proportion power calculation for binomial distribution (arcsine transformation)

              h = 0.3011521
              n = 231.7151
      sig.level = 0.05
          power = 0.9
    alternative = two.sided

NOTE: same sample sizes
```

FIGURE 10.3. Power analysis for proportions.

Based on these results, you would need to conduct a study with 232 students taking the novel course and 232 students receiving the existing course.

### Calculating Power for Chi-Square ($X^2$)

Power for $X^2$ is calculated with the **pwr.chisq.test()** function. The method for entering this in $R$ is:

```
>pwr.chisq.test(w=, n=, df=, sig.level=, power=)
```

In this formula, $w$ is the effect size and $n$ is the total sample size, as illustrated in Formula 10.3:

$$\text{Formula 10.3}: \quad w = \sum \frac{\sqrt{(p_o i - p_1 i)}}{p0_i}$$

Here, $p_0$ is the expected cell probabilities ($H_0$) and $p_1$ is the observed cell probabilities ($H_1$).

Effect size ($w$) can be calculated with the *pwr* function **ES.w2()**.

To illustrate this, let's consider the example we first discussed in Figure 9.4, which displays results on the relationship between cohabitation and discharge plans from County General Hospital. As displayed in Table 10.1, we need proportions for each cell to calculate the effect size.

TABLE 10.1. Proportional Outcomes for Hospital Patients

| Discharge Disposition | Not Cohabitating | Cohabitating |
|---|---|---|
| Home | .20 (*a*) | .34 (*b*) |
| Nursing | .20 (*c*) | .06 (*d*) |
| Rehab | .06 (*e*) | .14 (*f*) |

The cell proportions were calculated by dividing the total $n$ (50) by each cell's observed frequency. For example, in cell "a" for Home/Not Cohabiting, we calculated the table proportion as follows: $10/50 = 0.20$.

While you could hand-calculate this for each cell, you could use the **CrossTable()** function in the *gmodels* package to produce this for you as follows:

```
> CrossTable(hospital$discharge, hospital$cohabit,
prop.t=T, prop.r=F, prop.c=F, prop.chisq = F)
```

Notice that the results shown in Figure 10.4 are exactly as shown in our hand-calculated Table 10.1. In entering this function, we suppressed all the default options (row proportions, column proportions, and chi-square proportions) and only asked for the table proportions.

```
Cell Contents
|-------------------------|
|                       N |
|           N / Table Total |
|-------------------------|


Total Observations in Table:  50


            | cohabit
 discharge  |        No  |       Yes  | Row Total  |
------------|------------|------------|------------|
      Home  |        10  |        17  |        27  |
            |     0.200  |     0.340  |            |
------------|------------|------------|------------|
   Nursing  |        10  |         3  |        13  |
            |     0.200  |     0.060  |            |
------------|------------|------------|------------|
     Rehab  |         3  |         7  |        10  |
            |     0.060  |     0.140  |            |
------------|------------|------------|------------|
Column Total |       23  |        27  |        50  |
------------|------------|------------|------------|
```

FIGURE 10.4. CrossTable with table proportions only.

To calculate the needed effect size, $w$, we need to create a data matrix for the cell values using the following syntax:

```
p<-matrix(c(.20, .34, .20, .06, .06,
.14 ),byrow = T, nrow=3)
```

Typing **p** in the Console will display the matrix shown as follows:

```
       [,1] [,2]
[1,]   0.20 0.34
[2,]   0.20 0.06
[3,]   0.06 0.14
```

Notice the **matrix()** function includes the **byrow** and **nrow** options to emulate the shape of the previous tables. If the table had had four rows, then **nrow=4** would have been included.

Now you can calculate *w* with the following syntax:

```
>ES.w2(p)
```

The value of 0.3717058 for ES (*w*) appears in the Console.

The degrees of freedom (*df*) for this table is 2 as described previously in Chapter 9.

Assuming a significance level of 0.05 and a power of 0.9, the syntax to calculate sample size would be as follows:

```
>pwr.chisq.test(w= ES.w2(p), df=2, sig.level=.05,
power = .9)
```

The resulting output from the function is presented in Figure 10.5.

```
Chi squared power calculation

      w = 0.3717058
      N = 91.58557
     df = 2
sig.level = 0.05
  power = 0.9

NOTE: N is the number of observations
```

FIGURE 10.5. Power analysis for chi-square.

The results indicate that 92 participants would be needed to correctly detect a relationship between cohabitation and final disposition with the specified effect size, power, and significance level.

## Calculating Power for Correlation

Power for correlations is calculated with the **pwr.r.test()** function. The method for entering this in *R* is:

```
>pwr.r.test(r=, n=, sig.level=, power=, alternative=)
```

In this formula, *r* is the effect size measured by a linear correlation. The alternative option can be included to specify a two-tailed ("*two-sided*") or one-tailed ("*less*" or "*greater*") test.

As an example, consider a director of college admissions who wants to study the relationship between high school GPA and college GPA. The alternative hypothesis is that the correlation is greater than 0.27. The null hypothesis would be a correlation less than or equal to 0.27. The director sets the significance level to 0.05 and the power to 0.9. By utilizing the **pwr.r.test()** function, we can estimate how many students need to be included in our sample:

```
>pwr.r.test(r=.27, sig.level=.05, power=.9,
alternative="greater")
```

The output produced in the Console is displayed in Figure 10.6.

```
approximate correlation power calculation (arctangh transformation)

         n = 113.7318
         r = 0.27
 sig.level = 0.05
     power = 0.9
 alternative = greater
```

FIGURE 10.6. Power analysis for correlations.

A total of 114 participants are needed to be 90% certain of correctly detecting a correlation of this size between high school and college GPA.

## WHAT CAN YOU DO IF YOU DO NOT KNOW THE EFFECT SIZE?

In power analysis, the most difficult factor to estimate is the effect size (ES). As mentioned previously, you need to have some knowledge or data on the subject. It becomes extremely difficult to estimate ES when a study is new or innovative, however. In these cases, Cohen (1988) recommended benchmarks for small, medium, and large effect sizes for each of the statistical tests covered earlier, and these are displayed in Table 10.2.

TABLE 10.2. Cohen's Effect Size Benchmarks

| Statistical Test | Small | Medium | Large |
| --- | --- | --- | --- |
| *t*-Test | 0.20 | 0.50 | 0.80 |
| One-way ANOVA | 0.10 | 0.25 | 0.40 |
| Proportions | 0.20 | 0.50 | 0.80 |
| Chi-square | 0.10 | 0.30 | 0.50 |
| Correlation | 0.10 | 0.30 | 0.50 |

As previously noted, larger samples are required to detect small changes; therefore, it is often necessary to balance the need to detect change with the pragmatics of conducting research.

It is important to note that using the benchmarks shown in Table 10.2 may not be applicable to all disciplines (Kabacoff, 2011). An alternative approach would be to try different ranges of effect size, significance, and power. If you want to compare two groups and you are not certain of ES, you could produce a graph to plot ranges of effect size and corresponding sample size as shown in Figure 10.7 using the script available with this text entitled "Power analysis including sample sizes and ES."

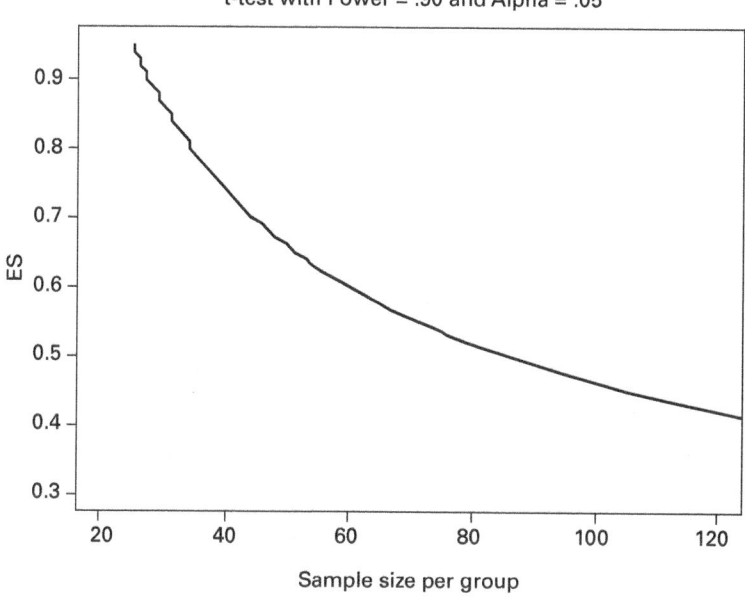

FIGURE 10.7. Example of a power curve of sample size.

As Figure 10.7 illustrates, as ES goes down (i.e., the ability to detect smaller changes), sample size increases. Using the script provided with this text, you can vary the power and alpha parameters.

The **plot.power.htest()** graphing function in the *pwr* package can be used to decide if there is any benefit to increasing power. The graph that results from this function displays how sample size increases as power increases.

As an example, let's see how the number of subjects would increase if power were increased in the example presented in the section on ANOVA earlier in the chapter. The syntax is shown as follows. Saving the results to a vector, which we called *g1*, is necessary so we can refer back to it later in our analysis:

```
>g1<-pwr.anova.test(k=4, f=.37, sig.level=.05,
power=.8)
>plot.power.htest(g1)
```

The results of entering this function in the Console are displayed in Figure 10.8.

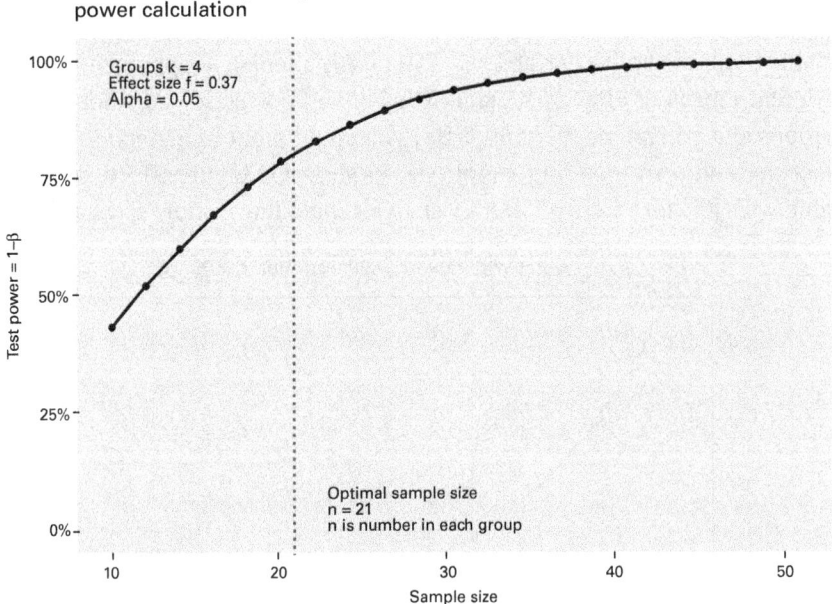

FIGURE 10.8. Example of power curve for one-way ANOVA.

The graph displays how the number of subjects would need to increase if power were increased. This can be useful in deciding if increasing power would be cost effective.

## CONCLUSION

Power analysis has become an important part of the design of research studies in establishing how many observations are necessary to be confident in your statistical analysis. If your sample is too small (i.e., your study is underpowered), you are less likely to reject the null when it should be rejected (i.e., correctly detect an existing relationship), and if you have more observations than needed, resources and time will be wasted.

## EXERCISES

Use the *pwr* package in *R* to answer each of the following questions:

1. You want to conduct a study designed to test an intervention to reduce test anxiety in college students. You decide to measure test anxiety using a scale

that ranges from a low of 21 to a high of 105. You need two groups: one to test the intervention and one that will get nothing and serve as a control group. Use $R$ to determine your sample size if you want to be able to detect a medium effect 90% of the time. As usual, $\alpha = 0.05$.

2. After working with your psychology professor all year, she offers you a summer job to work on a study. She wants to see if the number of years students were in school paid off financially. That is, is there a significant correlation between years in school and salary? She says that a power of 0.80 is required, and she has enough money in her budget to recruit 100 graduates. She asks you to do a power analysis to determine whether she would be able to detect small changes. How do you advise her?

a. Your professor decides to take another approach and determine what the effect size would be with her sample of 100. What is the effect size, and do you think it's worth conducting her study?

b. Finally, she asks you to do one more power analysis. With a sample of 100 graduates, what effect size could she detect with a power of 0.90?

3. The Office of Community Living at your university heard that you're a whiz at power analysis so they come to you with the following problem: Too many students are moving off campus after freshman year. Thirty-five percent move off campus, but the office would like to see that reduced to 20%. How many students would they need to recruit to study an intervention designed to keep students on campus? Since this is an important study, they'd like to be able to detect changes 90% of the time.

4. Your friend, who is studying education, is also trying to make use of your services! In the middle school where she is student teaching, the teachers are trying to design a study to determine what is the best way to deal with students cutting class. There are three options: (a) do what they're doing now—give detention after school to students who cut, (b) give students Saturday detention, or (c) have students attend an after-school program (not detention) to learn the importance of attending classes. How many middle schoolers should they put in each group to have a power of 0.80 and be able to detect medium effects?

5. Create a power plot like the one illustrated in Figure 10.8 with the analysis you did previously. Looking at the plot, estimate how much power you would have by increasing each sample size to 60. What power would you have if you could only recruit 40 students per group? What comments would you have?

# ADDITIONAL RESOURCES FOR LEARNING *R*

## SELECTED GENERAL BOOKS ON *R*

Auerbach, C., & Zeitlin, W. (2015). *Making your case: Using R for program evaluation*. New York, NY: Oxford University Press.

Dalgaard, P. (2008). *Introductory statistics with R* (2nd ed.). New York, NY: Springer.

de Vries, A., & Meys, J. (2015). *R for dummies* (2nd ed.). Hoboken, NJ: For Dummies.

Faraway, J. J. (2004). *Linear models with R*. Boca Raton, FL: Chapman and Hall/CRC.

Field, A., Miles, J., & Field, Z. (2012). *Discovering statistics using R*. Los Angeles, CA: Sage. Retrieved from http://studysites.sagepub.com/dsur/main.htm

Fox, J., Weisberg, S., & Fox, J. (2011). *An R companion to applied regression*. Thousand Oaks, CA: SAGE Publications.

Gandrud, C. (2015). *Reproducible research with R and R Studio* (2nd ed.). New York, NY: CRC Press. Retrieved from https://englianhu.files.wordpress.com/2016/01/reproducible-research-with-r-and-studio-2nd-edition.pdf

Kabacoff, R. I. (2011). *R in action: Data analysis and graphics with R* (2nd ed.). Shelter Island, NY: Manning Publications. Retrieved from https://github.com/kabacoff/RiA2

Lander, J. (2013). *R for everyone: Advanced analytics and graphics*. Upper Saddle River, NJ: Addison-Wesley Professional.

Teetor, P. (2011). *R cookbook*. Sebastopol, CA: O'Reilly Media.

## RESEARCH METHODS TEXTS

Begun, A. L., & Gregoire, T. K. (2014). *Conducting substance use research*. New York, NY: Oxford University Press.

Cohen, L., Manion, L., & Morrison, K. (2017). *Research methods in education* (8th ed.). New York, NY: Routledge.

Grinnell, R. M., & Unrau, Y. A. (2014). *Social work research and evaluation: Foundations of evidence-based practice* (10th ed.). New York, NY: Oxford University Press.

LoBiondo-Wood, G., & Haber, J. (2017). *Nursing research: Methods and critical appraisal for evidence-based practice* (9th ed.). St. Louis, MO: Mosby.

Mertens, D. M. (2014). *Research and evaluation in education and psychology: Integrating diversity with quantitative, qualitative, and mixed methods* (4th ed.). Thousand Oaks, CA: SAGE Publications.

Morling, B. (2017). *Research methods in psychology: Evaluating a world of information* (3rd ed.). New York, NY: W. W. Norton.

Roberts, A. R., & Yeager, K. (Eds.). (2006). *Foundations of evidence-based social work practice*. New York, NY: Oxford University Press.

Rubin, A., & Babbie, E. R. (2013). *Research methods for social work* (8th ed.). Belmont, CA: Brooks/Cole Publishing Company.

Shaughnessy, J. J., Zechmeister, E. B., & Zechmesiter, J. S. (2014). *Research methods in psychology* (10th ed.). Dubuque, IA: McGraw-Hill Education.

Vaughn, M. G., Pettus-Davis, C., & Shook, J. J. (2012). *Conducting research in juvenile and criminal justice settings*. New York, NY: Oxford University Press.

## WEBSITES

Dr. Auerbach's and Dr. Zeitlin's website: https://www.ssdanalysis.com

Institute for Digital Research and Education at UCLA: http://www.ats.ucla.edu/stat/r/

John Fox's website: http://socserv.mcmaster.ca/jfox/

Quick-R: http://www.statmethods.net/

R-Bloggers: http://www.r-bloggers.com/

R Meetups around the world: https://www.meetup.com/topics/r-project-for-statistical-computing/

The R Project for Statistical Computing: https://www.r-project.org

R Step-by-Step: http://www.indiana.edu/~phonlab/R/

UCLA Institute for Digital Research and Education: https://stats.idre.ucla.edu/r/

# MATH REVIEW

Introductory statistics students are sometimes reluctant because they do not feel competent in their mathematical abilities. Most introductory statistics courses do NOT require a terribly strong background in math. In fact, a short review of some basic mathematical concepts is likely all you need to be successful in your course if you feel a little weak in some areas.

KHAN ACADEMY® has excellent materials to review these concepts. To begin using these materials, we recommend that you simply enter the name of the video for the topic into the search bar on the KHAN ACADEMY® website. In most cases, you will find more than one video in a sequence. For example, if you type "Intro to Order of Operations" into the search bar, you will be taken to the first in a series of videos on the topic. We present multiple videos on a single topic because they may address different aspects of the topic or increase in complexity from one to the next.

Additionally, and importantly, you may also see opportunities to practice the skills taught in these videos by completing exercises in one or more places within the sequence. We strongly suggest that you attempt these exercises, as mastery of these will suggest that you are ready to move on to other more difficult topics. If you are having difficulty doing these, KHAN ACADEMY® embeds additional support within the exercises to use if they are needed.

Finally, some topics have reviews available for them. These are summaries of a topic without videos that often incorporate links to videos and some practice problems.

## BASIC ARITHMETIC

### Order of Operations Videos

- Intro to Order of Operations
- Order of Operations Example
- Worked Example: Order of Operations (PEMDAS)

## Sigma Notation Videos

- Summation Notation
- Converting Explicit Series Terms to Summation Notation
- Arithmetic Series in Sigma Notation
- Worked Example: Arithmetic Series (Sigma Notation)

## Scientific Notation Videos

- Introduction to Scientific Notation
- Scientific Notation Examples (there are two videos with this name)
- Scientific Notation Example: 0.0000000003457
- Scientific Notation Word Problem: Red Blood Cells
- Scientific Notation Word Problem: U.S. National Debt
- Scientific Notation Word Problem: Speed of Light

## Cancelling in Fractions Videos

- Intro to Fractions
- More About Fractions
- Intro to Equivalent Fractions
- Fractions in Lowest Terms
- Simplifying Complex Fractions

## BASIC ALGEBRA

- Same Thing to Both Sides of Equations
- One-Step Equation Intuitions
- Dividing Both Sides of an Equation
- One-Step Addition & Subtraction Equations
- One-Step Subtraction Equations
- One-Step Addition & Subtraction Equations: Fractions & Decimals
- Intro to Two-Step Equations
- Two-Step Equations Intuition
- Worked Example: Two-Step Equations

## UNDERSTANDING THE FORMULA OF A LINE

- Intro to Slope-Intercept Form
- Worked Examples: Slope-Intercept Intro
- Graph from Slope-Intercept Equation

- Slope-Intercept Equation From a Graph
- Slope-Intercept Equation From Slope & Point
- Slope-Intercept Equation From Two Points
- Slope-Intercept Form Problems

## FACTORIALS, COMBINATIONS, AND PERMUTATIONS VIDEOS

- Combination Formula
- Handshaking Combinations
- Combination Example: 9-Card Hands
- Factorial and Counting Seat Arrangements
- Permutation Formula
- Possible Three-Letter Words
- Zero Factorial or 0!
- Ways to Arrange Colors
- Intro to Combinations

# GETTING STARTED WITH *R*

## WHAT IS *R*?

*R* is an open-source, freely available statistical programming language and is compatible with Windows, OS X, Linux, and other UNIX variants. *R* is similar to *S*, a program developed at Bell Laboratories by John Chambers (Auerbach & Schudrich, 2013; The R Project for Statistical Computing, n.d.). Although *R* has been around since 1993, it has grown rapidly in popularity since 2010. It is a programming language for statistical analysis and graphics. The software offers the following features:

- an effective data handling and storage facility;
- a suite of operators for calculations on arrays, in particular matrices;
- a large, coherent, integrated collection of intermediate tools for data analysis;
- graphical facilities for data analysis and display either on-screen or on hardcopy; and
- a well-developed, simple and effective programming language, which includes conditionals, loops, user-defined recursive functions, and input and output facilities. (The R Project for Statistical Computing, n.d., paragraph 5)

In other words, *R* provides an environment where statistical techniques can be implemented (The R Project for Statistical Computing, n.d.). *R*'s capabilities have been extended through the development of functions and packages. Weisberg and Fox (2010) state, "One of the great strengths of R is that it allows users and experts in particular areas of statistics to add new capabilities to the software" (p. xiii).

For all of these reasons, we have begun working extensively in *R*, and we recommend that you do too!

To make working with *R* a bit easier, a number of freely available graphical user interfaces, or GUIs, have been developed. Among these are *RStudio, R Commander*, and *RKWard*. We use *RStudio*, as we have found it to be flexible and useful. The screen shots depicted throughout this book are based on our use of *RStudio*.

## INSTALLING *R* AND *RSTUDIO*

In this section you will learn how to install *R*, open *R* files, and enter *R* commands using the *RStudio* GUI.

Begin by downloading *R* and *RStudio* free of charge from links on the homepage of the Single-System Design Analysis website (http://www.ssdanalysis.com). If you are a Mac user, you will need to also install *XQuartz*. On this site you will also find videos on how to install the software. When you click the links for installing *R* and *RStudio*, you will be taken to external sites. Both *R* and *RStudio* are completely free and are considered safe and stable downloads.

Once these are installed, open *RStudio*. When you open it, your screen should look like Figure C.1.

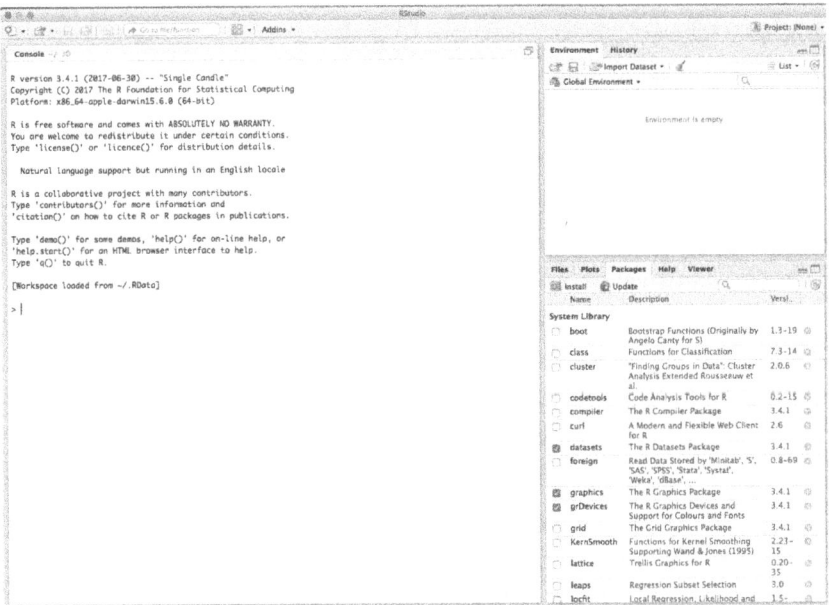

FIGURE C.1. A first look at *RStudio*.

## NAVIGATING *RSTUDIO*

The Console, located in the left pane of Figure C.1, is the area in which *R* functions are typed. After entering a function, pressing the <RETURN> key executes it. Pressing the up and down arrows enables you to scroll through functions in your history directly into the Console.

The top right pane contains three tabs: Environment, History, and Build. The Environment tab is where any files, known in *R* as data frames, you open or create during a session are listed along with vectors and variables. The History tab keeps a list of all *R* functions you enter. If you highlight a function in your history, you can then re-execute it by sending it to the Console. Your history is continuous from session to session and will not be cleared unless you clear it manually by clicking on the broom icon. The Build tab is used for programing *R* and will not be covered in this text.

The pane at the bottom right contains five tabs: Files, Plots, Packages, Help, and Viewer. The Files tab lists all files that are located in your default directory. The Plots tab opens a window that contains the most recent plot created during the session. Using the arrows in this tab helps you scroll through plots created during that session only. In this window there is an Export button that enables you to copy plots to the clipboard or save them in various formats such as a PDF, TIFF, or JPEG. The Help tab gives you access to *R* help files.

## SETTING YOUR WORKING DIRECTORY

It is good practice to begin your session by setting your default directory. To accomplish this, in the menu bar click on **Session/Set Working Directory/Choose Directory**. After you press <RETURN>, you will see the dialogue box presented in Figure C.2. Use this dialogue to navigate to the directory that contains the example files you downloaded for this book, and select Open.

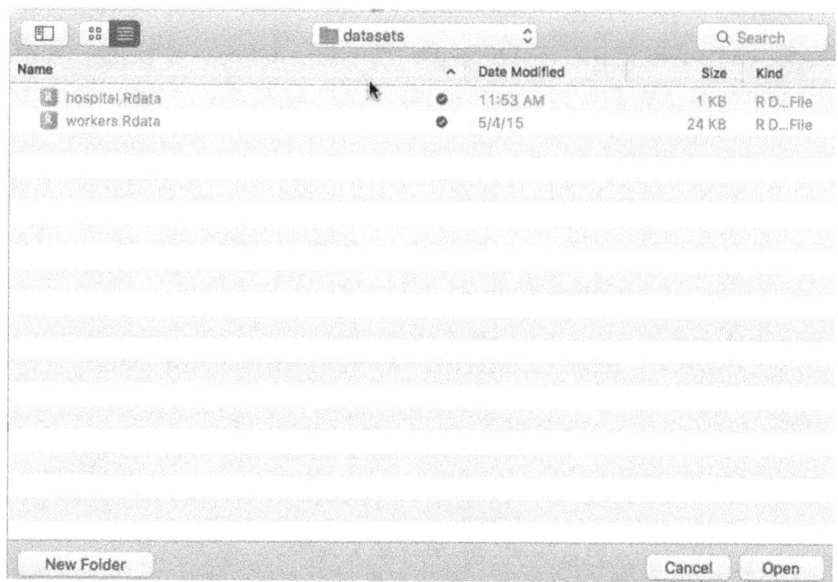

FIGURE C.2. Setting your working directory.

## OPENING A FILE

There are a number of methods for opening files in *RStudio*. The most common method is to employ the **File/Open File** menu choice located at the top of the menu bar of *RStudio*. As shown in Figure C.3, a dialogue box is presented similar to the one opened when the working directory was set. With this dialogue box, you can navigate to the directory containing files. Double-click the file *hospital. Rdata* to open it in *RStudio*. Notice, as displayed in Figure C.4, that *RStudio* queries you to click *Yes* to load the file into the global environment, which will complete the process.

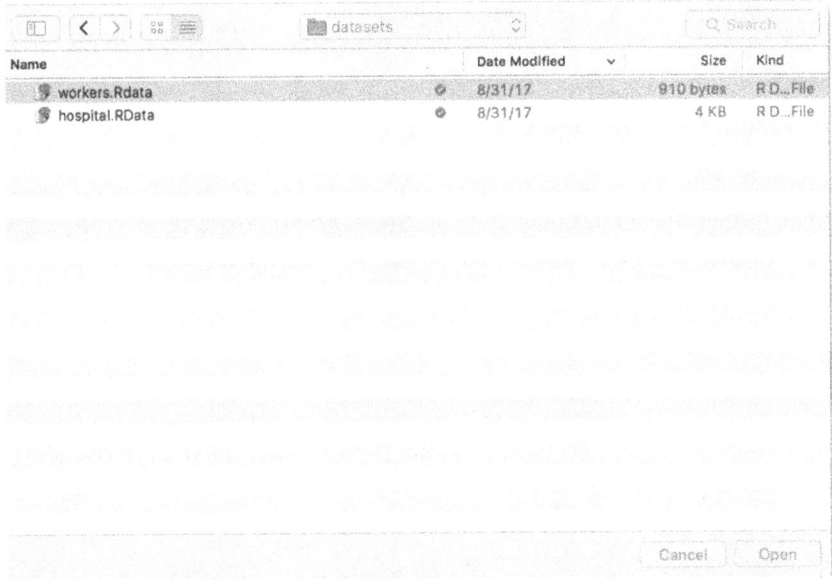

FIGURE C.3. Opening a file in *RStudio*.

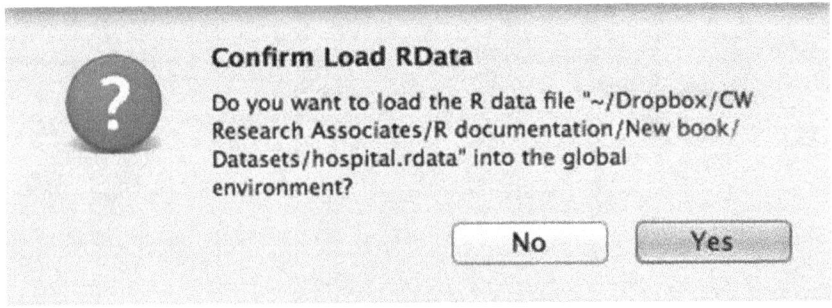

FIGURE C.4. Accepting your choice.

The *hospital* dataset is now listed in the top right *RStudio* pane. Alongside the *hospital* file are the number of observations, 50, and the number of variables, 19. Clicking on the spreadsheet icon to the right of the file in the Environment tab will display your data in a spreadsheet format in the upper left pane, as displayed in Figure C.5. When you do this, the Console will automatically drop into the lower left pane.

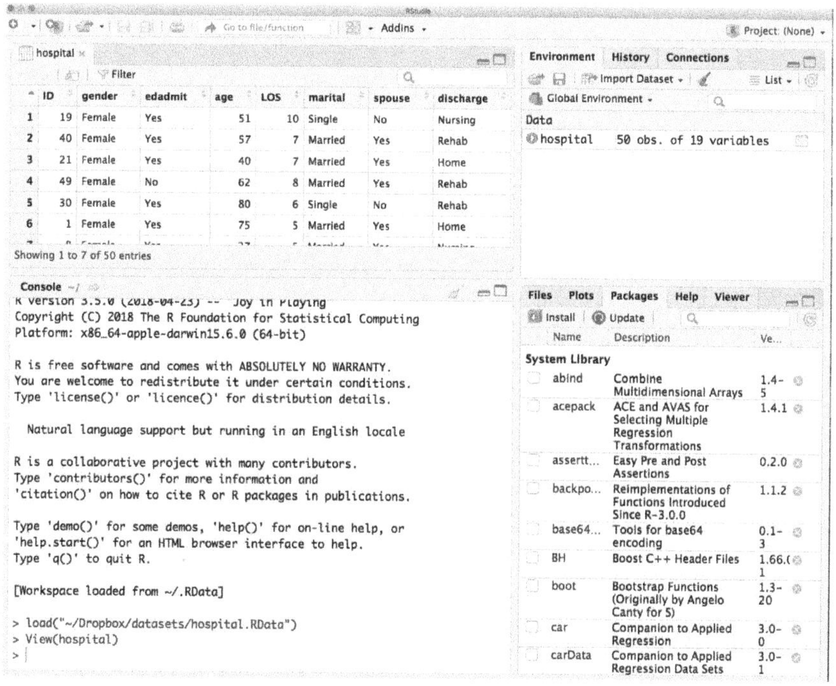

FIGURE C.5. Viewing your data in *RStudio*.

You cannot edit your data in this pane, but you can easily view it by scrolling left, right, up, or down. Additionally, simply grabbing the handles between the panes and stretching them or compressing them, as desired, can modify the size of each of these. Also notice that you can sort your entire dataset on any variable by simply clicking on the up or down arrows next to the variable name in gray.

As displayed in Figure C.6, you can also view the list of files in your working directory by clicking on the File tab in the bottom right pane. You can double-click on an *R* file (a file with the extension *.RData* or *.rdata*) to open it in *RStudio*.

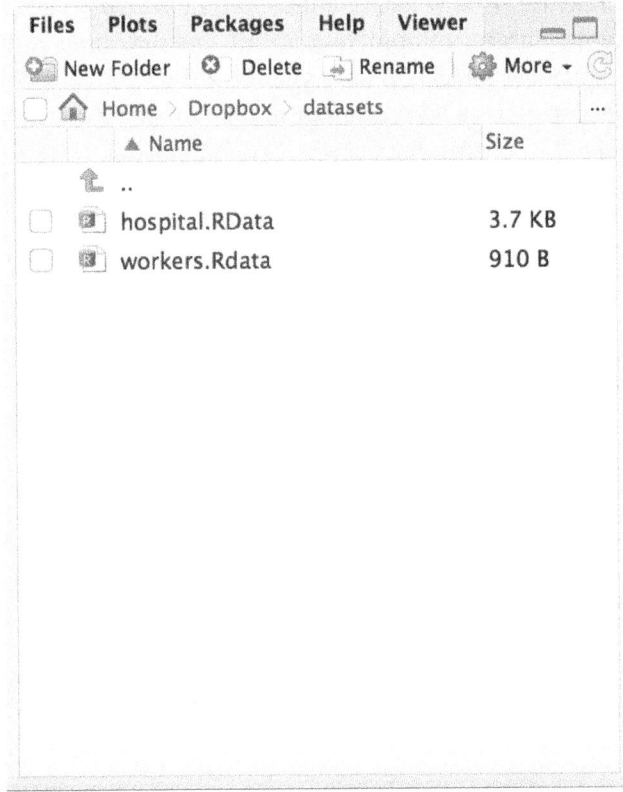

FIGURE C.6. Files listed in the Files pane.

## ENTERING YOUR FIRST *R* FUNCTION

Enter the following command into the Console in the bottom left pane of *RStudio:*

`>names(hospital)` and press <RETURN>.

You will obtain the results displayed in Figure C.7.

The `names()` function simply reports the names of variables contained in an *R* file.

## TO ATTACH OR NOT TO ATTACH? THAT IS THE QUESTION . . .

*R* can have multiple files entered into the environment at one time; however, you need a method to identify the variables from the file you want to analyze. The `attach()` function is one way that enables *R* to recognize the file in its search path so you can manipulate it. However, before opening a new file, you must remember to use the `detach()` function to remove it; otherwise, opening a different file with variables containing the same names as the current one will cause a conflict and an error message.

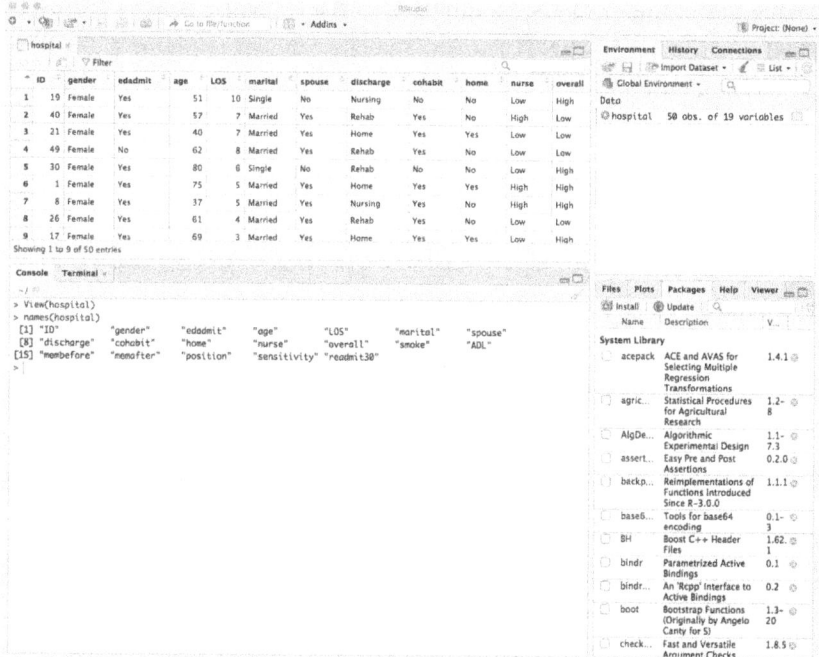

FIGURE C.7. Entering your first command: `names(hospital)`.

As a result, we generally recommend not using the **attach()** function to avoid this conflict. Instead, you can access variables in a file by using the *filename$* naming convention. Figure C.8 shows an example using this convention. Type the following:

```
>table(hospital$).
```

After you type *hospital$*, a pop-up list of all the variable names in the *hospital.Rdata* dataset will appear as displayed in Figure C.8. Select *marital* by clicking the variable name with your mouse.

Using the name of the file followed by a "$" prevents any potential conflicts and removes the need to attach your dataset.

FIGURE C.8. Using the filename$ convention.

You should have the following command in the Console: `table(hospital$ marital)`. Now press <ENTER>, and the table shown in Figure C.9 will be displayed in your Console.

```
> table(hospital$marital)

Divorced  Married   Single  Widowed
      11       23       12        4
>
```

FIGURE C.9. Table of marital status.

## ENDING YOUR SESSION

When you are ready to leave *RStudio*, end your session by simply clicking on **File/ Quit RStudio** in the menu bar. *RStudio* will then query you with the window displayed in Figure C.10. Since we do not care to save anything, click "Don't Save" and *RStudio* will close.

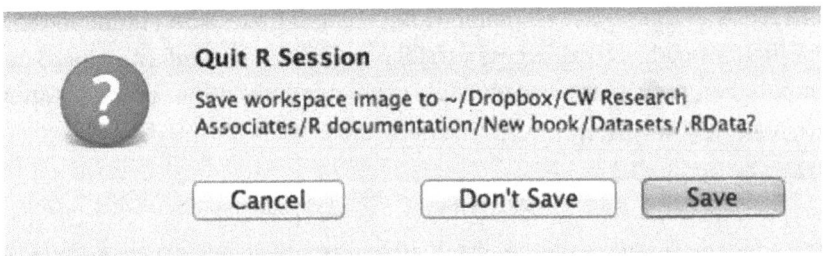

FIGURE C.10. Ending your *RStudio* session.

## PACKAGES

One of the appeals of *R* is the easily accessible collection of user-contributed packages. Currently, there over 11,000 packages on the Comprehensive R Archive Network (CRAN) written by thousands of user-developers (The R Project for Statistical Computing, n.d.). A package is simply a collection of prewritten *R* code to accomplish a particular task. For example, the *foreign* package allows users to import and transform files from other popular statistical packages, such as SPSS and Stata, to the *R* format. Another example is a package written by the authors, *SSDforR*, to analyze single-subject data.

It is likely that if a statistical method exists, there are one or more packages for it on CRAN. Once you open *RStudio*, you are connected to the world of CRAN and you can install any of the available packages easily and with no charge.

## Installing Packages

To install an *R* package, click on the Packages tab in the bottom right *RStudio* pane. Click on "Install" and the dialogue shown in Figure C.11 will be displayed. Make sure that "Repository (CRAN)" under "Install from" is selected. Later in the book you will be utilizing the *psych* and *Hmisc* packages. To install them now, type the following into the "Packages" dialogue and click Install: psych, Hmisc.

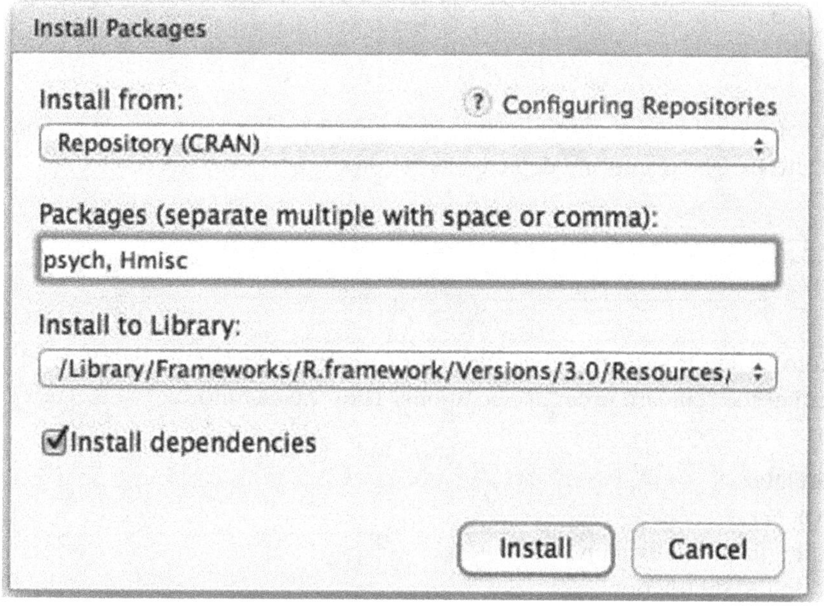

FIGURE C.11. Installing packages.

Packages only need to be installed once; however, to access them they must be required during each *R* session. The `require()` function can be utilized to enable a package. For example, `require(psych)` would allow you to access functions in the *psych* package. Alternatively, as displayed in Figure C.12, checking the box next to the package name in the Packages tab in the bottom right pane of *RStudio* also makes the package available for use.

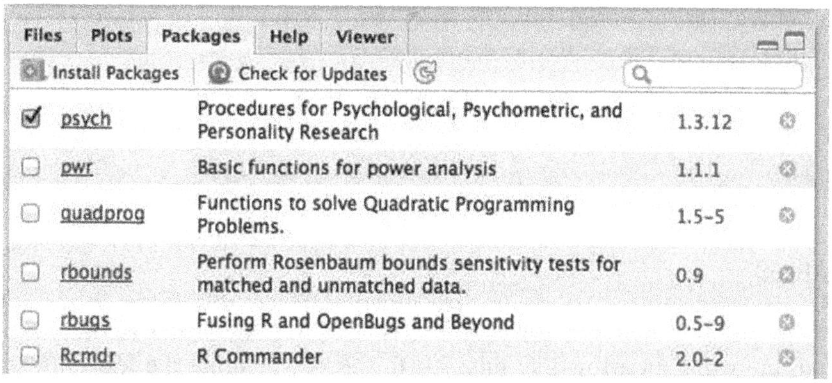

FIGURE C.12. Using the packages pane to require a package.

## SOME BASICS OF *R*

### *R* Can Do Math

Because *R* is a statistical programming language, it can also be used to perform basic mathematical functions. Entering 2 + 3 into the Console and pressing <RETURN> produces the following:

```
> 2 + 3
[1] 5
```

Now try your hand at multiplication by typing 3 * 4 into the Console and pressing <RETURN>. The results are depicted as follows:

```
> 3*4
[1] 12
```

More complex computations can be accomplished, but you will need to be mindful of the standard order of operations. They are as follows:

Parentheses
Exponents
Multiplication/Division
Addition/Subtraction

Operations inside parentheses take priority and are performed prior to any other process. For example, type the following into the Console:

```
> (20-10) / 2
[1] 5
```

In this case, the subtraction is performed first followed by division.

Exponents are entered into *R* with the ^ symbol. For example, try the following:

```
> 10^2
[1] 100
```

## VARIABLES

There are different methods for assigning values to variables. The most common methods are using an arrow like this: <- (formed by entering the less than symbol followed by a dash) or the = sign. You will obtain the same result using either method; however, the convention in *R* is to use <-. Type the following into the Console and press the <RETURN> key:

```
> x<-7
> x
[1] 7
```

You could repeat the same operation using the equal sign (=) to obtain the same result.

Now that *x* is stored in memory, it appears as a value in the Environment tab.

Be aware that *R* is case sensitive, so it differentiates between lowercase and uppercase variable names. Therefore, the variable *x* is not the same as the variable *X*. Also, variable names must start with an alphanumeric character. Furthermore, there cannot be any spaces between characters; however, the underscore (_) and dot (.) can be used to connect words. Special characters like the dash (-), asterisk (*), and slash (/) are not permissible as part of variable names.

You can remove a variable from memory using the **remove()** function. You can use the shortcut for the **remove()** command, **rm()**, to remove the variable *x* from memory. Simply type **rm(x)** into the Console and press <RETURN>. As shown, if **x** is typed into the Console after it is removed, the error presented as follows will appear. You will also notice that *x* was removed from the Environment tab:

```
> rm (x)
> x
Error: object 'x' not found
```

## TYPES OF VARIABLES

A variable in *R* can contain numbers, characters, or dates.

### Numeric Variables

Numeric variables are continuous variables that can be integers, both positive and negative, or decimals. We will re-create the *x* variable used in the previous section:

```
> x<-7
> x
[1] 7
```

A very useful function in *R* is **is.numeric()** that can be utilized to test if a variable is stored in *R* as a number. Try it out on the *x* variable by typing the following into the Console:

```
> is.numeric(x)
[1]  TRUE
```

For integers, *R* expects an "L" to be attached to the number. For example, type the following:

```
> y<-6L
> is.integer(y)
[1]  TRUE
```

It is also true that *y* is a numeric value, so using the **is.numeric**() function would also produce a result of "TRUE." Try the **class**() function:

```
>class(y)
[1] "integer"
```

### Character Variables

Although character string variables are nonmathematical, they are commonly used in data analysis. As displayed to follow, you can assign a character string to a variable:

```
x<-"hello"
>x
[1]  "hello"
```

As mentioned, *R* differentiates between upper- and lowercase characters; therefore, *R* would evaluate the same word in the following examples differently: *hello, Hello*, and *HELLO*.

### Dates

*R* contains a number of functions that provide for the manipulation of dates. A date can be directly entered employing the **as.Date**() function. An example is displayed to follow:

```
> admitted<-as.Date("2013-05-03")
> discharged<-as.Date("2013-05-23")
```

These dates represent when a patient was admitted to and discharged from a hospital. Notice that the dates were entered as a four-digit year, followed by a two-digit month and a two-digit day, all entered within quotation marks. This is the preferred method.

To calculate the total length of stay for the patient in days, the `as.numeric()` function can be utilized to convert a date into the number of days since January 1, 1970. With this function, the patient's length of stay in days can easily be calculated:

```
> discharged<-as.Date("2013-05-23")
> los<-as.numeric(discharged)-as.numeric(admitted)
> los
[1] 20
```

## Vectors

A vector is a collection of elements that can be stored as a variable. Vectors can be numbers, characters, dates, or any combination of these. The `c()`, or combine, command is a frequently used method to enter elements into a vector. For example, `x<-c(1, 2, 3, 4, 5)` creates a numeric vector named $x$ containing five elements, the numbers 1, 2, 3, 4, and 5.

$R$ is a vectorized programming language; any operation applied to a vector affects all the elements within it simultaneously. We can multiply our vector $x$ by a factor of 10. This example and its results are shown to follow. Notice that a comma separates each element. Remember to put the "c" in front of the open parenthesis:

```
> x<-c(1, 2, 3, 4, 5)
> x
[1] 1 2 3 4 5

> x<-x*10
> x
[1] 10 20 30 40 50
```

A vector can also contain characters like the following: `c("Tom", "Dick"," Harry")`. Each character element must be placed between quotation marks. The vector can be assigned to a variable as displayed here:

```
> y<-c("Tom", "Dick", "Harry")
> y

[1] "Tom" "Dick" "Harry"
```

A more complex example of the power of vectors is displayed to follow. Four patients admitted on different days are discharged from a hospital on the same day. Notice the **as.Date()** and **as.numeric()** functions are applied once to the vector *admitted* in the first and third steps, respectively:

```
> admitted<-as.Date(c("2013-12-20","2013-12-9","2013-12-
  11","2013-12-27"))
> discharged<-as.Date("2013-12-31")
> los<-as.numeric(discharged)-as.numeric(admitted)
> los
[1] 11 22 20 4
```

### Factor Variables

A factor variable is a type of categorical variable that can be represented as a string or a number. Converting categorical variables to factors has a number of advantages, especially when tables and graphics are used in data analysis. Factor variables are also useful in advanced statistical models like linear regression or logistic regression.

To illustrate, open the example dataset named *factor.Rdata*. To do this in *RStudio*, select **File** and then **Open File** and navigate to where the file is located. You will be prompted to load the file into *R*; select *Yes*. This data frame contains a single variable, *marital*. To look at the values of *marital*, use the **table()** function:

```
> table(factor$marital)

 1   2  3  4
16  95 44  4
```

Notice the *factor$* in the command before the variable name *marital*. As previously mentioned, a common variable such as *age* or *gender* can be present in multiple files you may be analyzing. Using the "filename$" convention in front of the variable name allows *R* to differentiate from which dataset you are selecting your variable and prevents any potential conflicts.

In your output, the first row represents the various categories of marital status and the second row represents the number of clients in each category. For example, we see that category 2 has 95 clients. The categories represent the following: 1 = single, 2 = married, 3 = widowed, and 4 = divorced, which you would need to know to interpret this table.

Any client who was single was entered as a 1, as a 2 if they were married, and so on. If *marital* were converted to a factor variable, the table would be more easily interpreted. The **factor()** function can be utilized to accomplish this. This is depicted as follows:

```
>f.marital<-factor(factor$marital,levels=c(1,2,3,4),
  labels=c("single","married","widowed","divorced"))
```

In the previous *R* function, the levels are defined using the `c()` function described previously. The `labels()` option is then used to assign labels to the categories in the order in which they are presented. Finally, a new vector/variable *f.marital* was created containing this factor information. To follow are the results of the `table()` function on the new factor variable. Note how this produces a more readable table:

```
> table(f.marital)

 single   married  widowed divorced
     16        95       44        4
```

In the section on data frames you will learn how to save a newly created variable to an existing file.

## MISSING VALUES

Missing responses are very common in social science research, particularly survey research. Respondents often decide not to answer a particular question on a survey and skip it or they choose not to respond to a question in an interview. *R* handles this by using "NA" to represent a missing response. To follow is an example that extends the previous example on hospital admission and discharge dates. Note that the third *admitted* date is missing and was entered into the *admitted* vector as "NA":

```
> admitted<-as.Date(c("2013-12-20","2013-12-9",NA,
  "2013-12-27"))
> discharged<-as.Date("2013-12-31")
> los<-as.numeric(discharged)-as.numeric(admitted)
> admitted
[1] "2013-12-20" "2013-12-09" NA             "2013-12-27"
> los

[1] 11 22 NA  4
```

In the fourth step, when *admitted* was entered into the Console, the displayed result contained the "NA" for the third patient. Finally, the number of days for *los* could not be calculated for this third patient, and an "NA" was assigned for this occurrence.

The `is.na()` function can be utilized to test for missing values. The use of this command is presented to follow. As indicated by the "TRUE," the third value is missing:

```
> is.na(los)
[1] FALSE FALSE TRUE FALSE
```

## DATA TRANSFORMATION

When analyzing data, there is often a need to modify or transform it into groups or to combine individual items in some way to form, for example, a scale.

The *hospital.Rdata* file will be used to illustrate some examples. If you do not already have this dataset open, in *RStudio* select **File** and then **Open File** from the menu bar and navigate to where you have saved your files. Double-click the file *hospital.Rdata*. When queried whether you want to load the file into the Global Environment, select *Yes*. You can now use the **names()** function to list the variable names in the file. This is displayed as follows:

```
>names(hospital)

 [1] "ID"        "gender"    "edadmit"     "age"        "LOS"
 [6] "marital"   "spouse"    "discharge"   "cohabit"    "home"
[11] "nurse"     "overall"   "smoke"       "ADL"        "membefore"
[16] "memafter"  "position"  "sensitivity" "readmit30"
```

### Recoding Data

Recoding is used to combine, collapse, or correct data. For example, the variable *age* is a numeric variable. In the *hospital* dataset, patients range in age from 18 to 100 years. For the purposes of analysis it may be helpful to collapse the data into the following categories: *less than 65, 65 to 69, 70 to 74, 75 to 79,* and *80 or older,* making it a categorical, or factor, variable. To do this recode you will need to use a number of *R's* logical operators presented in Table C.1.

TABLE C.1. Logical Operators

| Operator | Description |
| --- | --- |
| < | Less than |
| <= | Less than or equal to |
| > | Greater than |
| >= | Greater than or equal |
| == | Exactly equal to |
| != | Not equal to |
| !X | Not X |
| X \| Y | X or Y |
| X & Y | X and Y |
| isTRUE(x) | Test if x is true |

To recode the variable *age*, enter the following into the Console:

```
> agecat<-NA
> agecat[hospital$age < 65]<-0
> agecat[hospital$age >= 65 & hospital$age < 70]<-1
> agecat[hospital$age >= 70 & hospital$age < 75]<-2
> agecat[hospital$age >= 75 & hospital$age < 80]<-3
> agecat[hospital$age >= 80 ]<-4
```

The first statement creates a new variable called *agecat* and assigns missing values ("NA") to it initially as a default. The second statement assigns the value "0" to any observation whose age is (<) less than 65. This means a 0 is assigned to *agecat* for any case that has an *age* value less than 65 years. Similarly, the third statement assigns a value of "1" to *agecat* for any observation that has an *age* value between 65 and 69.9. The last three statements are similar.

Once you enter these, use the **table()** function to see the number of observations in each category. The results are displayed as follows:

```
> table(agecat)
agecat
 0  1  2  3  4
33  5  1  3  8
```

As mentioned earlier, it is more efficient to store a categorical variable as a factor variable. The syntax for doing this is displayed here:

```
>agecat<-factor(agecat,levels=c(0,1,2,3,4),labels=c
  ("< 65","65-69","70-74","75-79","80 or older"))
>table(agecat)
        agecat
            < 65     65-69     70-74     75-79 80 or older
              33         5         1         3           8
```

In this situation, *R* assigns numeric values sequentially to the factor variable so *<65 = 0, 65-69 = 1, 70-74 = 2, 75-79 = 3*, and *80 or older = 4*.

## Saving Your Transformations

Before the dataset can be saved, the new variables need to be added to the *hospital* file. As the following shows, the **data.frame()** function can be utilized to accomplish this:

```
>hospital1<-data.frame(hospital,agecat)
```

This command is appending the newly created variables to the *hospital* vector into a new vector called *hospital1*, which we are defining as a data frame. To save this vector you will first need to set your directory to the folder in which you have your datasets stored. To do this in *RStudio* select the desired working directory, as previously described. Now enter the following command:

```
>save(hospital1,file="hospital1.RData")
```

Alternately, you can check the box next to the newly created data frame in the Environment tab and then click on the disk icon. You will then be presented with a dialogue box. From there, you can navigate to where you would like the new file saved.

## SOME BASIC *R* COMMANDS

### Categorical Data

In the previous section, the `table()` command was used to describe categorical data. This function can also be used to display percentages and totals. Open the *hospital.Rdata* dataset by selecting **File/Open File** from the menu bar and navigate to where the file is located. Double-click the file to open it.

To begin, create the vector as follows:

```
> t.marital<-table(hospital$marital)
> t.marital

    Divorced  Married  Single  Widowed
        11       23      12        4
```

You can use the `prop.table()` function to display proportions. Notice that you need to have created a table vector first to do this:

```
> prop.table(t.marital)

Divorced  Married  Single  Widowed
  0.22     0.46    0.24     0.08
```

You can convert the proportions to percentages by multiplying by 100, as displayed here:

```
> prop.table(t.marital)*100

Divorced  Married  Single  Widowed
    22       46      24        8
```

As the following shows, the **addmargins()** function can be utilized to obtain totals:

```
> addmargins(t.marital)
    Divorced  Married   Single  Widowed      Sum
          11       23       12        4       50
```

## Numeric Data

Table C.2 displays a number of functions to describe numeric data.

TABLE C.2. Functions for Numeric Variables

| Function | Description |
|----------|-------------|
| mean(x) | Calculates the mean of x |
| median(x) | Calculates the median of x |
| sd(x) | Calculates the standard deviation of x |
| var(x) | Calculates the variance of x |
| range(x) | Calculates the range of x |
| sum(x) | Calculates the sum of x |
| min(x) | Displays the minimum value of x |
| max(x) | Displays the maximum value of x |

To follow is an example for calculating the mean of *age*:

```
> mean(hospital$age,na.rm=T)
[1] 58.58
```

Because *age* could have had some missing values, the **na.rm=T** argument is included in the statement. *R* returned a mean of "57.86." Because missing values are often present in data, it is preferable to include the missing value option.

To follow is an example of how to obtain a standard deviation:

```
> sd(hospital$age,na.rm=T)
[1] 16.42806
```

Here is an example of how to obtain a median:

```
> median(hospital$age,na.rm=T)
[1] 58.5
```

Typing each function to describe a variable can be tedious. The **summary()** command, displayed as follows, combines a number of calculated values. Notice that we do not need to include the missing values argument in this statement. Also notice the standard deviation is not included in the **summary()** output:

```
>summary(hospital$age)

Min. 1st Qu.  Median   Mean 3rd Qu.    Max.
 18.00   46.25   58.50   58.58   69.00  105.00
```

# DATASET EXPLANATIONS

## hospital.Rdata

The *hospital.Rdata* dataset contains information about 50 patients admitted to County General Hospital on one day. Some of the patients were admitted through the emergency department, and others were admitted directly by their physicians. The dataset contains the following variables:

| Variable | Description | Values | Variable Type |
|---|---|---|---|
| ID | Identification number | Unique numeric identifier for each patient | Numeric |
| gender | Patient gender | Female, Male | Factor |
| edadmit | Was the patient admitted to the hospital through the emergency department? | No, Yes | Factor |
| age | Age of patient | Actual age in years | Numeric |
| LOS | Length of stay for current admission in days | Actual number of days | Numeric |
| marital | Marital status of patient | Married, Single, Divorced, or Widowed | Factor |
| spouse | Does the patient have a spouse? | No, Yes | Factor |
| discharge | Where the patient went upon discharge | Home, Nursing = nursing home, Rehab = rehabilitation facility | Factor |
| cohabit | Does patient live with a spouse or partner? | No, Yes | Factor |

| Variable | Description | Values | Variable Type |
|---|---|---|---|
| home | Was the patient discharged home? | No, Yes | Factor |
| nurse | Patient satisfaction with nursing services | High, Low | Factor |
| overall | Patient overall satisfaction with experience in hospital | High, Low | Factor |
| smoke | Does patient currently smoke? | No, Yes | Factor |
| ADL | Activities of daily living—How well can patient care for him-/herself independently? | Scale ranges from 3 to 12 with higher numbers denoting more independence | Numeric |
| membefore | Patient's memory at admission | Scale ranges from 5 to 25 with higher numbers indicating better memory | Numeric |
| memafter | Patient's memory at discharge | Scale ranges from 5 to 25 with higher numbers indicating better memory | Numeric |
| position | Position of staff member most responsible for patient care | MD = attending physician, PA = physician assistant, RN = nurse | Factor |
| sensitivity | How much patient thought person in "position" was sensitive to patient's overall needs | Scale ranges from 1 to 20 with higher values indicating higher sensitivity | Numeric |
| readmit30 | Was patient readmitted to hospital within 30 days of last discharge? | No, Yes | Factor |

## worker.Rdata

The *worker.Rdata* dataset contains information about 30 employees of the Funky Fun Corporation. Funky Fun has new ownership, and the new owners initiated a series of workshops to improve administrative leadership within the organization. The employees included in the dataset have different jobs and were asked to complete a survey at two time points—before the workshops and again afterward. The idea was to assess how much the workshops impacted employees across the organization. The variables included in this dataset are as follows:

| Variable | Description | Values | Variable Type |
|---|---|---|---|
| ID | Identification number | Unique numeric identifier for each employee | Numeric |
| i1 | Have you thought about leaving the Funky Fun Corporation in the last year? | No, Yes | Factor |
| ambiguity | This is a measure of the degree of vagueness about what is expected of employees, measured before intervention | Measured on a 1–5 scale with higher scores representing clear expectations (i.e., less ambiguity) | Numeric |
| commun | A measure of satisfaction with communication at Funky Fun measured before intervention. | Measured on a 5-25 scale with higher scores representing higher levels of satisfaction with communication. | Numeric |
| totjobsat | Measure of overall satisfaction with employment at Funky Fun before intervention | Measured on a scale ranging from 60 to 150 with higher scores representing higher satisfaction | Numeric |
| position | Type of job held by subject | Manager, Worker, Clerical | Factor |
| age | Age of employee at first time point | Age in whole years | Numeric |
| ambiguity2 | This is a measure of the degree of vagueness about what is expected of employees, measured after intervention | Measured on a 1–5 scale with higher scores representing clear expectations (i.e., less ambiguity) | Numeric |

| Variable | Description | Values | Variable Type |
|---|---|---|---|
| commun2 | A measure of satisfaction with communication at Funky Fun, measured after intervention | Measured on a 5–25 scale with higher scores representing higher levels of satisfaction with communication | Numeric |
| totjobsat2 | Measure of overall satisfaction with employment at Funky Fun after intervention | Measured on a scale ranging from 60 to 150 with higher scores representing higher satisfaction | Numeric |

## STUDENTS

Twenty students living on campus at YOUR school have been recruited to participate in a number of interventions throughout the school year. This dataset is ONLY used for the examples with hand calculations so the data itself will be displayed in the table following the explanation of variables. An explanation of the variables is displayed in the following table:

| Variable | Description | Values |
|---|---|---|
| Resident | Identification number | Unique numeric identifier for each student |
| Dorm | The name of the dormitory that the student is living in this semester | Gryffindor, Hufflepuff, Ravenclaw, Slytherin |
| Gender | The gender with which the student most identifies | Female, Male |
| College | The school in which the student's major resides | Business, Humanities, Art & Science |
| Age | The age of the student on his/her last birthday | The actual age in years |
| Roommates | The number of roommates the student currently has | The number of actual roommates |
| GPA | The student's current grade point average | Scaled from 1 to 4 with 4 being the highest grades |
| Current credits | The number of credits the student is taking this semester | Actual number of credits |

| Variable | Description | Values |
|---|---|---|
| Courseload | The number of classes the student is taking this semester | Actual number of classes |
| Firstgen | Is this student the first in his/her family to go to college? | No, Yes |
| Accommodations | Does the student receive any accommodations because of a disability? | No, Yes |
| Stress | The student's average level of stress during the semester | Reported on a 1–10 scale with 1 being lowest stress and 10 being highest stress |
| Success | Self-perception of student in student role | Poor, Fair, Good, Excellent |
| Memory1 | Student memory before memory improvement workshop | High, Low |
| Memory2 | Student memory after memory improvement workshop | High, Low |
| WPM1 | Number of words student can read in one minute before speed-reading workshop | Actual number of words |
| WPM2 | Number of words student can read in one minute after speed-reading workshop | Actual number of words |

| Resident | Dorm | Gender | College | Age | Roommates | GPA | Current Credits | Course-load | Firstgen | Accommo-dations | Stress | Success | Memory1 | Memory2 | WPM1 | WPM2 |
|---|---|---|---|---|---|---|---|---|---|---|---|---|---|---|---|---|
| 1 | Gryffindor | Female | Humanities | 18 | 2 | 2.49 | 13 | 4 | Yes | No | 3 | Fair | Low | High | 110 | 140 |
| 2 | Hufflepuff | Male | Business | 18 | 0 | 3.66 | 12 | 6 | No | No | 10 | Excellent | High | High | 110 | 130 |
| 3 | Ravenclaw | Female | Art & Science | 21 | 1 | 3.43 | 12 | 4 | Yes | No | 5 | Excellent | Low | Low | 130 | 100 |
| 4 | Slytherin | Female | Art & Science | 25 | 1 | 2.77 | 18 | 6 | No | No | 2 | Good | Low | High | 120 | 150 |
| 5 | Gryffindor | Female | Art & Science | 18 | 0 | 2.96 | 14 | 4 | No | No | 10 | Poor | Low | Low | 130 | 110 |
| 6 | Hufflepuff | Female | Humanities | 20 | 1 | 1.49 | 13 | 4 | No | No | 5 | Poor | Low | Low | 130 | 120 |
| 7 | Ravenclaw | Female | Business | 19 | 3 | 2.61 | 14 | 4 | No | Yes | 10 | Fair | Low | High | 150 | 140 |
| 8 | Slytherin | Female | Humanities | 22 | 3 | 3.95 | 16 | 5 | No | No | 6 | Excellent | Low | High | 120 | 110 |
| 9 | Gryffindor | Female | Art & Science | 18 | 2 | 2.85 | 16 | 5 | No | No | 7 | Good | Low | High | 130 | 150 |
| 10 | Hufflepuff | Female | Humanities | 21 | 2 | 1.17 | 12 | 3 | Yes | Yes | 1 | Poor | High | High | 140 | 110 |
| 11 | Ravenclaw | Male | Art & Science | 20 | 2 | 2.99 | 14 | 4 | No | No | 2 | Excellent | High | High | 120 | 130 |
| 12 | Slytherin | Male | Business | 21 | 3 | 2.62 | 15 | 5 | No | No | 3 | Fair | Low | Low | 110 | 140 |
| 13 | Gryffindor | Male | Business | 19 | 2 | 3.1 | 14 | 4 | No | Yes | 8 | Good | Low | Low | 130 | 130 |
| 14 | Hufflepuff | Male | Business | 20 | 2 | 3.97 | 12 | 4 | No | No | 4 | Excellent | High | Low | 150 | 170 |

| 15 | Ravenclaw | Male | Business | 21 | 2 | 1.03 | 13 | 4 | Yes | No | 6 | Poor | Low | Low | 120 | 120 |
| 16 | Slytherin | Male | Humanities | 20 | 1 | 2.44 | 18 | 6 | Yes | No | 4 | Fair | Low | High | 95 | 160 |
| 17 | Gryffindor | Female | Art & Science | 19 | 2 | 2.46 | 12 | 4 | Yes | Yes | 3 | Poor | High | High | 120 | 110 |
| 18 | Hufflepuff | Male | Business | 19 | 2 | 3.24 | 18 | 6 | Yes | No | 2 | Good | High | High | 140 | 140 |
| 19 | Ravenclaw | Male | Humanities | 20 | 3 | 3.12 | 14 | 4 | No | No | 8 | Good | Low | Low | 120 | 110 |
| 20 | Slytherin | Male | Business | 21 | 3 | 4 | 13 | 4 | No | No | 10 | Excellent | Low | High | 90 | 140 |

# GLOSSARY

## A

**Alternate hypothesis**—A statement used in hypothesis testing that is contrary to the null hypothesis. It is a prediction of how the researcher thinks that at least two variables are related in the population, and it is the hypothesis that is accepted if the null hypothesis can be rejected.

## B

**Bar plot**—A graph that represents data with spaced rectangular bars. The length of the bar denotes the frequency or proportion with which the data occurs.

**Bin**—A way of sorting continuous data in a histogram that splits data into consecutive discrete intervals.

## C

**Central limit theorem**—Theorem that states that if repeated samples are taken from any distribution, the shape of that distribution, the sampling distribution, approaches normality as the number of samples taken from the original distribution increases. Additionally, the sample mean of the sampling distribution will equal the mean of the population, and the variance will be the variance of the population divided by the sample size.

**Coefficient of determination**—Defined as $r^2$, the coefficient of determination describes the proportion of the dependent variable that is explained by the regression model.

**Combinations**—A selection of all or part of a set in which the order of the selection does not matter.

**Confidence intervals**—The estimated range within which it is calculated that a true population parameter lies. Confidence intervals are described in terms of how confident one can be that the true population parameter lies in the specified range.

**Constant**—Having a fixed value; something that does not vary.

**Contingency table**—Also referred to as a cross-tabulation, this is a table showing how the frequency and/or percentage of two categorical variables occur in a given set of data.

**Covariance**—A measure of how two variables move together without regard to units of change. Closely related to correlations, positive covariances indicate that as one variable increases, the other does as well, while negative covariances indicate that as one variable increases, the other decreases.

**Criterion of significance**—A statement of the likelihood of rejecting the null hypothesis when it is true (i.e., a Type I error).

# D

**Dependence**—A condition where two variables are not independent.
**Dependent variable**—The variable that is measured in a study and that is the outcome of interest. In an experiment, this variable is not manipulated directly by the researcher and its value is thought to be influenced by one or more independent variables.

# E

**Effect size**—A description of the magnitude of change or difference between groups.
**Expected value**—A predicted value of a variable.

# H

**Heterogeneity**—The extent to which populations, observations, or results differ. The opposite of homogeneity.
**Histogram**—A graph that depicts a frequency or density distribution of a continuous variable. A histogram groups observations in bins that are continuous ranges of observed data.
**Homogeneity**—The degree to which populations, observations, or results are similar. The opposite of heterogeneity.
**Hypothesis**—An idea that can be tested through research. Most frequently this term is synonymous with the alternate hypothesis.
**Hypothesis testing**—A process by which one attempts to answer a hypothesis. The result is that one can reject the null hypothesis and, therefore, accept the alternate or that one is unable to accept the null.

# I

**Independence**—The idea that one event in no way affects the probability of another event occurring.
**Independent variable**—A variable that is measured in a study that is not contingent upon any other variable. Often referred to as a predictor, these are the variables that are manipulated and thought to have an impact on the dependent variable.
**Indicator**—A measure of a current condition or variable.
**Interval**—A measure of the distance between attributes of a variable.

# K

**Kurtosis**—A measure of symmetry. A normal distribution has kurtosis of three.

# M

**Mean**—The average of a set of numbers.

**Median**—The middle number in a value-ordered set of numbers. The median is the 50th percentile of the ordered set.

**Mode**—The value that occurs the most in a set of numbers.

**Mutually exclusive**—The idea that two or more events cannot occur simultaneously.

# N

**Nominal**—A level of measurement for categorical variables that is unordered.

**Nonprobability sample**—A type of sample in which all elements in the population do not have an equal chance of being selected for inclusion.

**Normal distribution**—A class of distributions defined by a specific formula. This distribution is symmetrical and bell shaped with the mean, median, and mode of the distribution being equal.

**Null hypothesis**—A hypothesis of no difference or the opposite of an alternate that one attempts to accept over an alternate hypothesis.

# O

**Observation**—A measured piece of data. Also referred to as a case.

**Odds**—The ratio of the probability that an event will occur to the probability that it will not occur.

**Ordinal**—Categorical variables that are classified by a rank order.

**Outlier**—An observation point that is detached from the main body of observations. Also referred to as an extreme value.

# P

**Parameter**—A value (e.g., mean, percent, standard deviation) associated with a population.

**Permutation**—A selection of all or part of a set in which the order of the selection does matter.

**Power**—The probability of being able to detect a difference between two groups if one existed in the population and if the research was replicated with the same sample size. If there were a power of 0.9, this would tell us that, given the same sample size, a statistically significant difference between groups would be detected 90% of the time.

**Power analysis**—A technique utilized to detect the probability of not detecting a statistically significant difference when one exists.

**Probability sample**—A type of sample in which all elements in the population have an equal chance of being selected for inclusion. A common type of probability sample utilizes random selection. This method helps ensure that the various sample units are representative of the population from which they have been drawn.

# R

**Residual**—The difference between an observed value and a predicted value.

# S

**Sample**—A subset of a population.

**Sampling bias**—The systematic error that causes some members of a population to be less likely or more likely to be included in a sample. This can result in a sample that is not representative of a larger population.

**Sampling distribution**—A theoretical probability distribution that represents that statistic's value if a large number of samples of a given size were drawn from the population. These are used in inferential statistics to determine how unusual an observation or calculated value is.

**Scatter plot**—Set of points displayed on a graph to show how two continuous variables are related to one another. Each point indicates the value of both the variable plotted on the x-axis and the variable plotted on the y-axis for a given observation.

**Skewness**—A measure of the direction and symmetry of a distribution. A distribution that is perfectly symmetrical has a skewness of zero.

**Standard error of the estimate**—An estimate of the accuracy with which a sample represents the population it was drawn from.

**Standard error of the mean**—A measure of how much a sample mean deviates from the actual mean.

**Standard normal distribution**—A special case of the normal distribution in which the mean of the distribution is zero and the standard deviation is one.

**Statistic**—A value (e.g., mean, percent, standard deviation) associated with a sample.

**Statistical significance**—The likelihood of making a Type I error.


# T

**Type I error**—Drawing the erroneous conclusion that we think something occurs in a population when it actually does not.

**Type II error**—Drawing the erroneous conclusion that we do not think something occurs in a population when it actually does.


# V

**Variable**—a concept that can be measured and can differ between observations. Concepts that are not variable are constant.


# Z

*z*-**Score**—A transformed score that indicates how many standard deviations a score lies from the mean when the distribution has a mean of zero and a standard deviation of one.

# TABLES OF DISTRIBUTIONS

TABLE F.1. Area Under the Standard Normal Curve

Table of Standard Normal Probabilities for Negative Z-scores

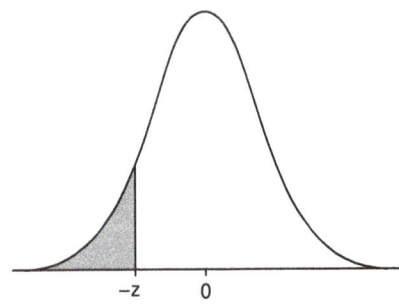

| z | 0.00 | 0.01 | 0.02 | 0.03 | 0.04 | 0.05 | 0.06 | 0.07 | 0.08 | 0.09 |
|---|------|------|------|------|------|------|------|------|------|------|
| −3.4 | 0.0003 | 0.0003 | 0.0003 | 0.0003 | 0.003 | 0.0003 | 0.0003 | 0.0003 | 0.0003 | 0.0002 |
| −3.3 | 0.0005 | 0.0005 | 0.0005 | 0.0004 | 0.0004 | 0.0004 | 0.0004 | 0.0004 | 0.0004 | 0.0003 |
| −3.2 | 0.0007 | 0.0007 | 0.0006 | 0.0006 | 0.0006 | 0.0006 | 0.0006 | 0.0005 | 0.0005 | 0.0005 |
| −3.1 | 0.0010 | 0.0009 | 0.0009 | 0.0009 | 0.0008 | 0.0008 | 0.0008 | 0.0008 | 0.0007 | 0.0007 |
| −3.0 | 0.0013 | 0.0013 | 0.0013 | 0.0012 | 0.0012 | 0.0011 | 0.0011 | 0.0011 | 0.0010 | 0.0010 |
| −2.9 | 0.0019 | 0.0018 | 0.0018 | 0.0017 | 0.0016 | 0.0016 | 0.0015 | 0.0015 | 0.0014 | 0.0014 |
| −2.8 | 0.0026 | 0.0025 | 0.0024 | 0.0023 | 0.0023 | 0.0022 | 0.0021 | 0.0021 | 0.0020 | 0.0019 |
| −2.7 | 0.0035 | 0.0034 | 0.0033 | 0.0032 | 0.0031 | 0.0030 | 0.0029 | 0.0020 | 0.0027 | 0.0026 |
| −2.6 | 0.0047 | 0.0045 | 0.0044 | 0.0043 | 0.0041 | 0.0040 | 0.0039 | 0.0038 | 0.0037 | 0.0036 |
| −2.5 | 0.0062 | 0.0060 | 0.0059 | 0.0057 | 0.0055 | 0.0054 | 0.0052 | 0.0051 | 0.0049 | 0.0048 |
| −2.4 | 0.0082 | 0.0080 | 0.0078 | 0.0075 | 0.0073 | 0.0071 | 0.0069 | 0.0068 | 0.0066 | 0.0064 |
| −2.3 | 0.0107 | 0.0104 | 0.0102 | 0.0099 | 0.0096 | 0.0094 | 0.0091 | 0.0089 | 0.0087 | 0.0084 |
| −2.2 | 0.0139 | 0.0136 | 0.0132 | 0.0129 | 0.0125 | 0.0122 | 0.0119 | 0.0116 | 0.0113 | 0.0110 |
| −2.1 | 0.0179 | 0.0174 | 0.0170 | 0.0166 | 0.0162 | 0.0158 | 0.0154 | 0.0150 | 0.0146 | 0.0143 |
| −2.0 | 0.0228 | 0.0222 | 0.0217 | 0.0212 | 0.0207 | 0.0202 | 0.0197 | 0.0192 | 0.0188 | 0.0183 |
| −1.9 | 0.0287 | 0.0281 | 0.0274 | 0.0268 | 0.0262 | 0.0256 | 0.0250 | 0.0244 | 0.0239 | 0.0233 |
| −1.8 | 0.0359 | 0.0351 | 0.0344 | 0.0336 | 0.0329 | 0.0322 | 0.0314 | 0.0307 | 0.0301 | 0.0294 |
| −1.7 | 0.0446 | 0.0436 | 0.0427 | 0.0418 | 0.0409 | 0.0401 | 0.0392 | 0.0384 | 0.0375 | 0.0367 |
| −1.6 | 0.0548 | 0.0537 | 0.0526 | 0.0516 | 0.0505 | 0.0495 | 0.0485 | 0.0475 | 0.0465 | 0.0455 |
| −1.5 | 0.0668 | 0.0655 | 0.0643 | 0.0630 | 0.0618 | 0.0606 | 0.0594 | 0.0582 | 0.0571 | 0.0559 |
| −1.4 | 0.0808 | 0.0793 | 0.0778 | 0.0764 | 0.0749 | 0.0735 | 0.0721 | 0.0708 | 0.0694 | 0.0681 |
| −1.3 | 0.0968 | 0.0951 | 0.0934 | 0.0918 | 0.0901 | 0.0885 | 0.0869 | 0.0853 | 0.0838 | 0.0823 |
| −1.2 | 0.1151 | 0.1131 | 0.1112 | 0.1093 | 0.1075 | 0.1056 | 0.1038 | 0.1020 | 0.1003 | 0.0985 |

| z | 0.00 | 0.01 | 0.02 | 0.03 | 0.04 | 0.05 | 0.06 | 0.07 | 0.08 | 0.09 |
|---|---|---|---|---|---|---|---|---|---|---|
| −1.1 | 0.1357 | 0.1335 | 0.1314 | 0.1292 | 0.1271 | 0.1251 | 0.1230 | 0.1210 | 0.1190 | 0.1170 |
| −1.0 | 0.1587 | 0.1562 | 0.1539 | 0.1515 | 0.1492 | 0.1469 | 0.1446 | 0.1423 | 0.1401 | 0.1379 |
| −0.9 | 0.1841 | 0.1814 | 0.1788 | 0.1762 | 0.1736 | 0.1711 | 0.1685 | 0.1660 | 0.1635 | 0.1611 |
| −0.8 | 0.2119 | 0.2090 | 0.2061 | 0.2033 | 0.2005 | 0.1977 | 0.1949 | 0.1922 | 0.1894 | 0.1867 |
| −0.7 | 0.2420 | 0.2389 | 0.2358 | 0.2327 | 0.2296 | 0.2266 | 0.2236 | 0.2206 | 0.2177 | 0.2148 |
| −0.6 | 0.2743 | 0.2709 | 0.2676 | 0.2643 | 0.2611 | 0.2578 | 0.2546 | 0.2514 | 0.2483 | 0.2451 |
| −0.5 | 0.3085 | 0.3050 | 0.3015 | 0.2981 | 0.2946 | 0.2912 | 0.2877 | 0.2843 | 0.2810 | 0.2776 |
| −0.4 | 0.3446 | 0.3409 | 0.3372 | 0.3336 | 0.3300 | 0.3264 | 0.3228 | 0.3192 | 0.3156 | 0.3121 |
| −0.3 | 0.3821 | 0.3783 | 0.3745 | 0.3707 | 0.3669 | 0.3632 | 0.3594 | 0.3557 | 0.3520 | 0.3483 |
| −0.2 | 0.4207 | 0.4168 | 0.4129 | 0.4090 | 0.4052 | 0.4013 | 0.3974 | 0.3936 | 0.3897 | 0.3859 |
| −0.1 | 0.4602 | 0.4562 | 0.4522 | 0.4483 | 0.4443 | 0.4404 | 0.4364 | 0.4325 | 0.4286 | 0.4247 |
| −0.0 | 0.5000 | 0.4960 | 0.4920 | 0.4880 | 0.4840 | 0.4801 | 0.4761 | 0.4721 | 0.4681 | 0.4641 |

Table of Standard Normal Probabilities for Positive Z-scores

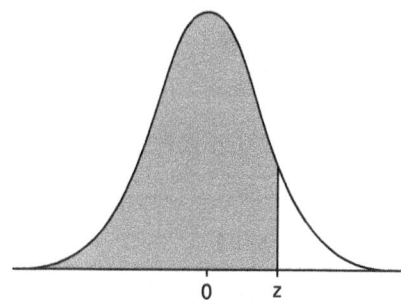

| z | 0.00 | 0.01 | 0.02 | 0.03 | 0.04 | 0.05 | 0.06 | 0.07 | 0.08 | 0.09 |
|---|---|---|---|---|---|---|---|---|---|---|
| 0.0 | 0.5000 | 0.5040 | 0.5080 | 0.5120 | 0.5160 | 0.5199 | 0.5239 | 0.5279 | 0.5319 | 0.5359 |
| 0.1 | 0.5398 | 0.5438 | 0.5478 | 0.5517 | 0.5557 | 0.5596 | 0.5636 | 0.5675 | 0.5714 | 0.5753 |
| 0.2 | 0.5793 | 0.5832 | 0.5871 | 0.5910 | 0.5948 | 0.5987 | 0.6026 | 0.6064 | 0.6103 | 0.6141 |
| 0.3 | 0.6179 | 0.6217 | 0.6255 | 0.6293 | 0.6331 | 0.6368 | 0.6406 | 0.6443 | 0.6480 | 0.6517 |
| 0.4 | 0.6554 | 0.6591 | 0.6628 | 0.6664 | 0.6700 | 0.6736 | 0.6772 | 0.6808 | 0.6844 | 0.6879 |
| 0.5 | 0.6915 | 0.6950 | 0.6985 | 0.7019 | 0.7054 | 0.7088 | 0.7123 | 0.7157 | 0.7190 | 0.7224 |
| 0.6 | 0.7257 | 0.7291 | 0.7324 | 0.7357 | 0.7389 | 0.7422 | 0.7454 | 0.7486 | 0.7517 | 0.7549 |
| 0.7 | 0.7580 | 0.7611 | 0.7642 | 0.7673 | 0.7704 | 0.7734 | 0.7764 | 0.7794 | 0.7823 | 0.7852 |

TABLE F.1. Continued

| z | 0.00 | 0.01 | 0.02 | 0.03 | 0.04 | 0.05 | 0.06 | 0.07 | 0.08 | 0.09 |
|---|------|------|------|------|------|------|------|------|------|------|
| 0.8 | 0.7881 | 0.7910 | 0.7939 | 0.7967 | 0.7995 | 0.8023 | 0.8051 | 0.8078 | 0.8106 | 0.8133 |
| 0.9 | 0.8159 | 0.8186 | 0.8212 | 0.8238 | 0.8264 | 0.8289 | 0.8315 | 0.8340 | 0.8365 | 0.8389 |
| 1.0 | 0.8413 | 0.8438 | 0.8461 | 0.8485 | 0.8508 | 0.8531 | 0.8554 | 0.8577 | 0.8599 | 0.8621 |
| 1.1 | 0.8643 | 0.8665 | 0.8686 | 0.8708 | 0.8729 | 0.8749 | 0.8770 | 0.8790 | 0.8810 | 0.8830 |
| 1.2 | 0.8849 | 0.8869 | 0.8888 | 0.8907 | 0.8925 | 0.8944 | 0.8962 | 0.8980 | 0.8997 | 0.9015 |
| 1.3 | 0.9032 | 0.9049 | 0.9066 | 0.9082 | 0.9099 | 0.9115 | 0.9131 | 0.9147 | 0.9162 | 0.9177 |
| 1.4 | 0.9192 | 0.9207 | 0.9222 | 0.9236 | 0.9251 | 0.9265 | 0.9279 | 0.9292 | 0.9306 | 0.9319 |
| 1.5 | 0.9332 | 0.9345 | 0.9357 | 0.9370 | 0.9382 | 0.9394 | 0.9406 | 0.9418 | 0.9429 | 0.9441 |
| 1.6 | 0.9452 | 0.9463 | 0.9474 | 0.9484 | 0.9495 | 0.9505 | 0.9515 | 0.9525 | 0.9535 | 0.9545 |
| 1.7 | 0.9554 | 0.9564 | 0.9573 | 0.9582 | 0.9591 | 0.9599 | 0.9608 | 0.9616 | 0.9625 | 0.9633 |
| 1.8 | 0.9641 | 0.9649 | 0.9656 | 0.9664 | 0.9671 | 0.9678 | 0.9686 | 0.9693 | 0.9699 | 0.9706 |
| 1.9 | 0.9713 | 0.9719 | 0.9726 | 0.9732 | 0.9738 | 0.9744 | 0.9750 | 0.9756 | 0.9761 | 0.9767 |
| 2.0 | 0.9772 | 0.9778 | 0.9783 | 0.9788 | 0.9793 | 0.9798 | 0.9803 | 0.9808 | 0.9812 | 0.9817 |
| 2.1 | 0.9821 | 0.9826 | 0.9830 | 0.9834 | 0.9838 | 0.9878 | 0.9846 | 0.9850 | 0.9854 | 0.9857 |
| 2.2 | 0.9861 | 0.9864 | 0.9868 | 0.9871 | 0.9875 | 0.9878 | 0.9881 | 0.9884 | 0.9887 | 0.9890 |
| 2.3 | 0.9893 | 0.9896 | 0.9898 | 0.9901 | 0.9904 | 0.9906 | 0.9909 | 0.9911 | 0.9913 | 0.9916 |
| 2.4 | 0.9918 | 0.9920 | 0.9922 | 0.9925 | 0.9927 | 0.9929 | 0.9931 | 0.9932 | 0.9934 | 0.9936 |
| 2.5 | 0.9938 | 0.9940 | 0.9941 | 0.9943 | 0.9945 | 0.9946 | 0.9948 | 0.9949 | 0.9951 | 0.9952 |
| 2.6 | 0.9953 | 0.9955 | 0.9956 | 0.9957 | 0.9959 | 0.9960 | 0.9961 | 0.9962 | 0.9963 | 0.9964 |
| 2.7 | 0.9965 | 0.9966 | 0.9967 | 0.9968 | 0.9969 | 0.9970 | 0.9971 | 0.9972 | 0.9973 | 0.9974 |
| 2.8 | 0.9974 | 0.9975 | 0.9976 | 0.9977 | 0.9977 | 0.9978 | 0.9979 | 0.9979 | 0.9980 | 0.9981 |
| 2.9 | 0.9981 | 0.9982 | 0.9982 | 0.9983 | 0.9984 | 0.9984 | 0.9985 | 0.9985 | 0.9986 | 0.9986 |
| 3.0 | 0.9987 | 0.9987 | 0.9987 | 0.9988 | 0.9988 | 0.9989 | 0.9989 | 0.9989 | 0.9990 | 0.9990 |
| 3.1 | 0.9990 | 0.9991 | 0.9991 | 0.9991 | 0.9992 | 0.9992 | 0.9992 | 0.9992 | 0.9993 | 0.9993 |
| 3.2 | 0.9993 | 0.9993 | 0.9994 | 0.9994 | 0.9994 | 0.9994 | 0.9994 | 0.9995 | 0.9995 | 0.9995 |
| 3.3 | 0.9995 | 0.9995 | 0.9995 | 0.9996 | 0.9996 | 0.9996 | 0.9996 | 0.9996 | 0.9996 | 0.9997 |
| 3.4 | 0.9997 | 0.9997 | 0.9997 | 0.9997 | 0.9997 | 0.9997 | 0.9997 | 0.9997 | 0.9997 | 0.9998 |

Note that the probabilities given in this table represent the area to the LEFT of the z-score.

The area to the RIGHT of a z-score = 1 - the area to the LEFT of the z-score

TABLE F.2. The Student's Table of Critical *t*-Values

For a One-Tailed Test:

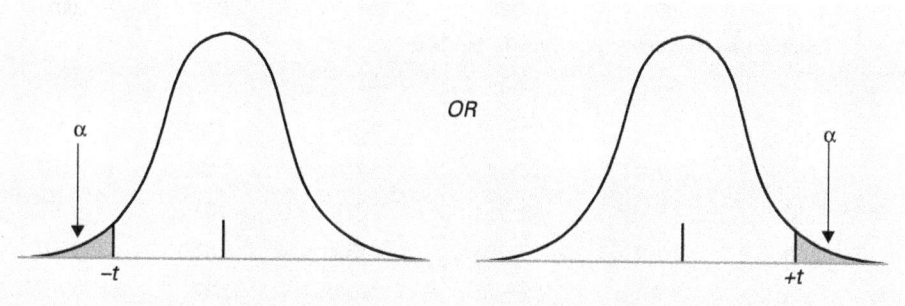

For a Two-Tailed Test:

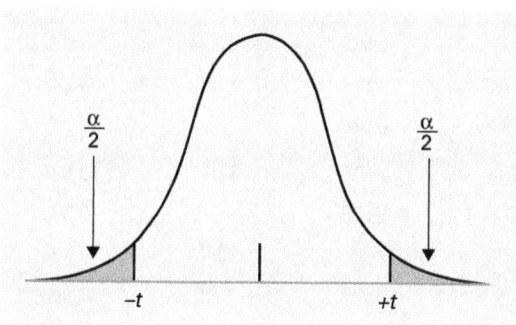

| | Level of Significance (α) for One-Tailed Test | | | | | |
|---|---|---|---|---|---|---|
| | .1 | .05 | .025 | .01 | .005 | .001 |
| | Level of Significance (α) for Two-Tailed Test | | | | | |
| df | .2 | .1 | .05 | .02 | .01 | .002 |
| 1 | 3.078 | 6.314 | 12.706 | 31.821 | 63.657 | 318.313 |
| 2 | 1.886 | 2.920 | 4.303 | 6.965 | 9.925 | 22.327 |
| 3 | 1.638 | 2.353 | 3.182 | 4.541 | 5.841 | 10.215 |
| 4 | 1.533 | 2.132 | 2.776 | 3.747 | 4.604 | 7.173 |
| 5 | 1.476 | 2.015 | 2.571 | 3.365 | 4.032 | 5.893 |
| 6 | 1.440 | 1.943 | 2.447 | 3.143 | 3.707 | 5.208 |
| 7 | 1.415 | 1.895 | 2.365 | 2.998 | 3.499 | 4.782 |
| 8 | 1.397 | 1.860 | 2.306 | 2.896 | 3.355 | 4.499 |
| 9 | 1.383 | 1.833 | 2.262 | 2.821 | 3.250 | 4.296 |

TABLE F.2. Continued

| df | Level of Significance ($\alpha$) for One-Tailed Test | | | | | |
|---|---|---|---|---|---|---|
| | .1 | .05 | .025 | .01 | .005 | .001 |
| | Level of Significance ($\alpha$) for Two-Tailed Test | | | | | |
| | .2 | .1 | .05 | .02 | .01 | .002 |
| 10 | 1.372 | 1.812 | 2.228 | 2.764 | 3.169 | 4.143 |
| 11 | 1.363 | 1.796 | 2.201 | 2.718 | 3.106 | 4.024 |
| 12 | 1.356 | 1.782 | 2.179 | 2.681 | 3.055 | 3.929 |
| 13 | 1.350 | 1.771 | 2.160 | 2.650 | 3.012 | 3.852 |
| 14 | 1.345 | 1.761 | 2.145 | 2.624 | 2.977 | 3.787 |
| 15 | 1.341 | 1.753 | 2.131 | 2.602 | 2.947 | 3.733 |
| 16 | 1.337 | 1.746 | 2.120 | 2.583 | 2.921 | 3.686 |
| 17 | 1.333 | 1.740 | 2.110 | 2.567 | 2.898 | 3.646 |
| 18 | 1.330 | 1.734 | 2.101 | 2.552 | 2.878 | 3.610 |
| 19 | 1.328 | 1.729 | 2.093 | 2.539 | 2.861 | 3.579 |
| 20 | 1.325 | 1.725 | 2.086 | 2.528 | 2.845 | 3.552 |
| 21 | 1.323 | 1.721 | 2.080 | 2.518 | 2.831 | 3.527 |
| 22 | 1.321 | 1.717 | 2.074 | 2.508 | 2.819 | 3.505 |
| 23 | 1.319 | 1.714 | 2.069 | 2.500 | 2.807 | 3.485 |
| 24 | 1.318 | 1.711 | 2.064 | 2.492 | 2.797 | 3.467 |
| 25 | 1.316 | 1.708 | 2.060 | 2.485 | 2.787 | 3.450 |
| 26 | 1.315 | 1.706 | 2.056 | 2.479 | 2.779 | 3.435 |
| 27 | 1.314 | 1.703 | 2.052 | 2.473 | 2.771 | 3.421 |
| 28 | 1.313 | 1.701 | 2.048 | 2.467 | 2.763 | 3.408 |
| 29 | 1.311 | 1.699 | 2.045 | 2.462 | 2.756 | 3.396 |
| 30 | 1.310 | 1.697 | 2.042 | 2.457 | 2.750 | 3.385 |
| 31 | 1.309 | 1.696 | 2.040 | 2.453 | 2.744 | 3.375 |
| 32 | 1.309 | 1.694 | 2.037 | 2.449 | 2.738 | 3.365 |
| 33 | 1.308 | 1.692 | 2.035 | 2.445 | 2.733 | 3.356 |
| 34 | 1.307 | 1.691 | 2.032 | 2.441 | 2.728 | 3.348 |
| 35 | 1.306 | 1.690 | 2.030 | 2.438 | 2.724 | 3.340 |
| 36 | 1.306 | 1.688 | 2.028 | 2.434 | 2.719 | 3.333 |
| 37 | 1.305 | 1.687 | 2.026 | 2.431 | 2.715 | 3.326 |
| 38 | 1.304 | 1.686 | 2.024 | 2.429 | 2.712 | 3.319 |
| 39 | 1.304 | 1.685 | 2.023 | 2.426 | 2.708 | 3.313 |
| 40 | 1.303 | 1.684 | 2.021 | 2.423 | 2.704 | 3.307 |

TABLE F.2. Continued

| df | Level of Significance (α) for One-Tailed Test | | | | | |
|---|---|---|---|---|---|---|
| | .1 | .05 | .025 | .01 | .005 | .001 |
| | Level of Significance (α) for Two-Tailed Test | | | | | |
| | .2 | .1 | .05 | .02 | .01 | .002 |
| 41 | 1.303 | 1.683 | 2.020 | 2.421 | 2.701 | 3.301 |
| 42 | 1.302 | 1.682 | 2.018 | 2.418 | 2.698 | 3.296 |
| 43 | 1.302 | 1.681 | 2.017 | 2.416 | 2.695 | 3.291 |
| 44 | 1.301 | 1.680 | 2.015 | 2.414 | 2.692 | 3.286 |
| 45 | 1.301 | 1.679 | 2.014 | 2.412 | 2.690 | 3.281 |
| 46 | 1.300 | 1.679 | 2.013 | 2.410 | 2.687 | 3.277 |
| 47 | 1.300 | 1.678 | 2.012 | 2.408 | 2.685 | 3.273 |
| 48 | 1.299 | 1.677 | 2.011 | 2.407 | 2.682 | 3.269 |
| 49 | 1.299 | 1.677 | 2.010 | 2.405 | 2.680 | 3.265 |
| 50 | 1.299 | 1.676 | 2.009 | 2.403 | 2.678 | 3.261 |
| 51 | 1.298 | 1.675 | 2.008 | 2.402 | 2.676 | 3.258 |
| 52 | 1.298 | 1.675 | 2.007 | 2.400 | 2.674 | 3.255 |
| 53 | 1.298 | 1.674 | 2.006 | 2.399 | 2.672 | 3.251 |
| 54 | 1.297 | 1.674 | 2.005 | 2.397 | 2.670 | 3.248 |
| 55 | 1.297 | 1.673 | 2.004 | 2.396 | 2.668 | 3.245 |
| 56 | 1.297 | 1.673 | 2.003 | 2.395 | 2.667 | 3.242 |
| 57 | 1.297 | 1.672 | 2.002 | 2.394 | 2.665 | 3.239 |
| 58 | 1.296 | 1.672 | 2.002 | 2.392 | 2.663 | 3.237 |
| 59 | 1.296 | 1.671 | 2.001 | 2.391 | 2.662 | 3.234 |
| 60 | 1.296 | 1.671 | 2.000 | 2.390 | 2.660 | 3.232 |
| 61 | 1.296 | 1.670 | 2.000 | 2.389 | 2.659 | 3.229 |
| 62 | 1.295 | 1.670 | 1.999 | 2.388 | 2.657 | 3.227 |
| 63 | 1.295 | 1.669 | 1.998 | 2.387 | 2.656 | 3.225 |
| 64 | 1.295 | 1.669 | 1.998 | 2.386 | 2.655 | 3.223 |
| 65 | 1.295 | 1.669 | 1.997 | 2.385 | 2.654 | 3.220 |
| 66 | 1.295 | 1.668 | 1.997 | 2.384 | 2.652 | 3.218 |
| 67 | 1.294 | 1.668 | 1.996 | 2.383 | 2.651 | 3.216 |
| 68 | 1.294 | 1.668 | 1.995 | 2.382 | 2.650 | 3.214 |
| 69 | 1.294 | 1.667 | 1.995 | 2.382 | 2.649 | 3.213 |
| 70 | 1.294 | 1.667 | 1.994 | 2.381 | 2.648 | 3.211 |

TABLE F.2. Continued

| df | Level of Significance (α) for One-Tailed Test | | | | | |
|---|---|---|---|---|---|---|
| | .1 | .05 | .025 | .01 | .005 | .001 |
| | Level of Significance (α) for Two-Tailed Test | | | | | |
| | .2 | .1 | .05 | .02 | .01 | .002 |
| 71 | 1.294 | 1.667 | 1.994 | 2.380 | 2.647 | 3.209 |
| 72 | 1.293 | 1.666 | 1.993 | 2.379 | 2.646 | 3.207 |
| 73 | 1.293 | 1.666 | 1.993 | 2.379 | 2.645 | 3.206 |
| 74 | 1.293 | 1.666 | 1.993 | 2.378 | 2.644 | 3.204 |
| 75 | 1.293 | 1.665 | 1.992 | 2.377 | 2.643 | 3.202 |
| 76 | 1.293 | 1.665 | 1.992 | 2.376 | 2.642 | 3.201 |
| 77 | 1.293 | 1.665 | 1.991 | 2.376 | 2.641 | 3.199 |
| 78 | 1.292 | 1.665 | 1.991 | 2.375 | 2.640 | 3.198 |
| 79 | 1.292 | 1.664 | 1.990 | 2.374 | 2.640 | 3.197 |
| 80 | 1.292 | 1.664 | 1.990 | 2.374 | 2.639 | 3.195 |
| 81 | 1.292 | 1.664 | 1.990 | 2.373 | 2.638 | 3.194 |
| 82 | 1.292 | 1.664 | 1.989 | 2.373 | 2.637 | 3.193 |
| 83 | 1.292 | 1.663 | 1.989 | 2.372 | 2.636 | 3.191 |
| 84 | 1.292 | 1.663 | 1.989 | 2.372 | 2.636 | 3.190 |
| 85 | 1.292 | 1.663 | 1.988 | 2.371 | 2.635 | 3.189 |
| 86 | 1.291 | 1.663 | 1.988 | 2.370 | 2.634 | 3.188 |
| 87 | 1.291 | 1.663 | 1.988 | 2.370 | 2.634 | 3.187 |
| 88 | 1.291 | 1.662 | 1.987 | 2.369 | 2.633 | 3.185 |
| 89 | 1.291 | 1.662 | 1.987 | 2.369 | 2.632 | 3.184 |
| 90 | 1.291 | 1.662 | 1.987 | 2.368 | 2.632 | 3.183 |
| 91 | 1.291 | 1.662 | 1.986 | 2.368 | 2.631 | 3.182 |
| 92 | 1.291 | 1.662 | 1.986 | 2.368 | 2.630 | 3.181 |
| 93 | 1.291 | 1.661 | 1.986 | 2.367 | 2.630 | 3.180 |
| 94 | 1.291 | 1.661 | 1.986 | 2.367 | 2.629 | 3.179 |
| 95 | 1.291 | 1.661 | 1.985 | 2.366 | 2.629 | 3.178 |
| 96 | 1.290 | 1.661 | 1.985 | 2.366 | 2.628 | 3.177 |
| 97 | 1.290 | 1.661 | 1.985 | 2.365 | 2.627 | 3.176 |
| 98 | 1.290 | 1.661 | 1.984 | 2.365 | 2.627 | 3.175 |
| 99 | 1.290 | 1.660 | 1.984 | 2.365 | 2.626 | 3.175 |

TABLE F.2. Continued

| df | Level of Significance ($\alpha$) for One-Tailed Test | | | | | |
|---|---|---|---|---|---|---|
| | .1 | .05 | .025 | .01 | .005 | .001 |
| | Level of Significance ($\alpha$) for Two-Tailed Test | | | | | |
| | .2 | .1 | .05 | .02 | .01 | .002 |
| 100 | 1.290 | 1.660 | 1.984 | 2.364 | 2.626 | 3.174 |
| 120 | 1.289 | 1.658 | 1.980 | 2.358 | 2.617 | 3.373 |
| ∞ | 1.282 | 1.645 | 1.960 | 2.326 | 2.576 | 3.090 |

TABLE F.3. Critical *F*-Values

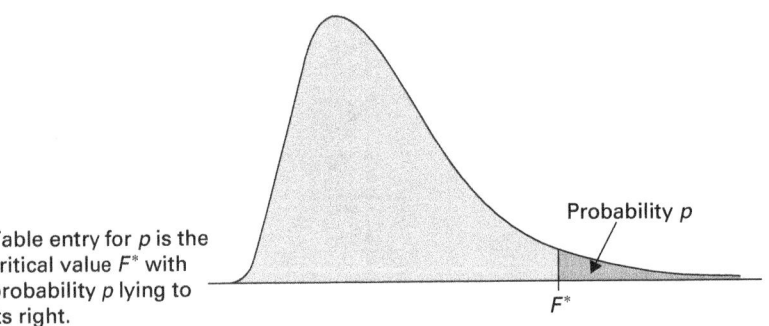

Table entry for *p* is the critical value *F*\* with probability *p* lying to its right.

Probability *p*

*F*\*

| | | **Degrees of freedom in the numerator** | | | | | | | | |
|---|---|---|---|---|---|---|---|---|---|---|
| | *p* | 1 | 2 | 3 | 4 | 5 | 6 | 7 | 8 | 9 |
| 1 | .050 | 161.45 | 199.50 | 215.71 | 224.58 | 230.16 | 233.99 | 236.77 | 238.88 | 240.54 |
| | .010 | 4052.2 | 4999.5 | 5403.4 | 5624.6 | 5763.6 | 5859.0 | 5928.4 | 5981.1 | 6022.5 |
| 2 | .050 | 18.51 | 19.00 | 19.16 | 19.25 | 19.30 | 19.33 | 19.35 | 19.37 | 19.38 |
| | .010 | 98.50 | 99.00 | 99.17 | 99.25 | 99.30 | 99.33 | 99.36 | 99.37 | 99.39 |
| 3 | .050 | 10.13 | 9.55 | 9.28 | 9.12 | 9.01 | 8.94 | 8.89 | 8.85 | 8.81 |
| | .010 | 34.12 | 30.82 | 29.46 | 28.71 | 28.24 | 27.91 | 27.67 | 27.49 | 27.35 |
| 4 | .050 | 7.71 | 6.94 | 6.59 | 6.39 | 6.26 | 6.16 | 6.09 | 6.04 | 6.00 |
| | .010 | 21.20 | 18.00 | 16.69 | 15.98 | 15.52 | 15.21 | 14.98 | 14.80 | 14.66 |
| 5 | .050 | 6.61 | 5.79 | 5.41 | 5.19 | 5.05 | 4.95 | 4.88 | 4.82 | 4.77 |
| | .010 | 16.26 | 13.27 | 12.06 | 11.39 | 10.97 | 10.67 | 10.46 | 10.29 | 10.16 |
| 6 | .050 | 5.99 | 5.14 | 4.76 | 4.53 | 4.39 | 4.28 | 4.21 | 4.15 | 4.10 |
| | .010 | 13.75 | 10.92 | 9.78 | 9.15 | 8.75 | 8.47 | 8.26 | 8.10 | 7.98 |
| 7 | .050 | 5.59 | 4.74 | 4.35 | 4.12 | 3.97 | 3.87 | 3.79 | 3.73 | 3.68 |
| | .010 | 12.25 | 9.55 | 8.45 | 7.85 | 7.46 | 7.19 | 6.99 | 6.84 | 6.72 |
| 8 | .050 | 5.32 | 4.46 | 4.07 | 3.84 | 3.69 | 3.58 | 3.50 | 3.44 | 3.39 |
| | .010 | 11.26 | 8.65 | 7.59 | 7.01 | 6.63 | 6.37 | 6.18 | 6.03 | 5.91 |
| 9 | .050 | 5.12 | 4.26 | 3.86 | 3.63 | 3.48 | 3.37 | 3.29 | 3.23 | 3.18 |
| | .010 | 10.56 | 8.02 | 6.99 | 6.42 | 6.06 | 5.80 | 5.61 | 5.47 | 5.35 |
| 10 | .050 | 4.96 | 4.10 | 3.71 | 3.48 | 3.33 | 3.22 | 3.14 | 3.07 | 3.02 |
| | .010 | 10.04 | 7.56 | 6.55 | 5.99 | 5.64 | 5.39 | 5.20 | 5.06 | 4.94 |
| 11 | .050 | 4.84 | 3.98 | 3.59 | 3.36 | 3.20 | 3.09 | 3.01 | 2.95 | 2.90 |
| | .010 | 9.65 | 7.21 | 6.22 | 5.67 | 5.32 | 5.07 | 4.89 | 4.74 | 4.63 |
| 12 | .050 | 4.75 | 3.89 | 3.49 | 3.26 | 3.11 | 3.00 | 2.91 | 2.85 | 2.80 |
| | .010 | 9.33 | 6.93 | 5.95 | 5.41 | 5.06 | 4.82 | 4.64 | 4.50 | 4.39 |
| 13 | .050 | 4.67 | 3.81 | 3.41 | 3.18 | 3.03 | 2.92 | 2.83 | 2.77 | 2.71 |
| | .010 | 9.07 | 6.70 | 5.74 | 5.21 | 4.86 | 4.62 | 4.44 | 4.30 | 4.19 |

Degrees of freedom in the denominator

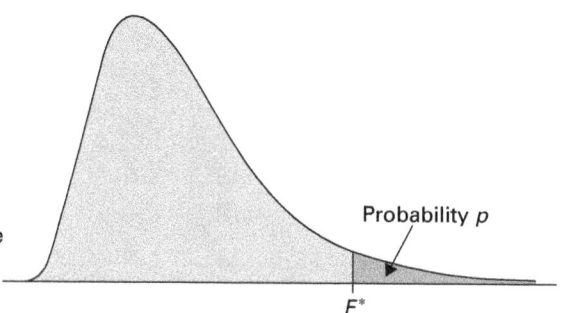

Table entry for $p$ is the critical value $F^*$ with probability $p$ lying to its right.

Probability $p$

$F^*$

| Degrees of freedom in the numerator | | | | | | | | | | |
|---|---|---|---|---|---|---|---|---|---|---|
| **10** | **12** | **15** | **20** | **25** | **30** | **40** | **50** | **60** | **120** | **1000** |
| 241.88 | 243.91 | 245.95 | 248.01 | 249.26 | 250.10 | 251.14 | 251.77 | 252.20 | 253.25 | 254.19 |
| 6055.8 | 6106.3 | 6157.3 | 6208.7 | 6239.8 | 6260.6 | 6286.8 | 6302.5 | 6313.0 | 6339.4 | 6362.7 |
| 19.40 | 19.41 | 19.43 | 19.45 | 19.46 | 19.46 | 19.47 | 19.48 | 19.48 | 19.49 | 19.49 |
| 99.40 | 99.42 | 99.43 | 99.45 | 99.46 | 99.47 | 99.47 | 99.48 | 99.48 | 99.49 | 99.50 |
| 8.79 | 8.74 | 8.70 | 8.66 | 8.63 | 8.62 | 8.59 | 8.58 | 8.57 | 8.55 | 8.53 |
| 27.23 | 27.05 | 26.87 | 26.69 | 26.58 | 26.50 | 26.41 | 26.35 | 26.32 | 26.22 | 26.14 |
| 5.96 | 5.91 | 5.86 | 5.80 | 5.77 | 5.75 | 5.72 | 5.70 | 5.69 | 5.66 | 5.63 |
| 14.55 | 14.37 | 14.20 | 14.02 | 13.91 | 13.84 | 13.75 | 13.69 | 13.65 | 13.56 | 13.47 |
| 4.74 | 4.68 | 4.62 | 4.56 | 4.52 | 4.50 | 4.46 | 4.44 | 4.43 | 4.40 | 4.37 |
| 10.05 | 9.89 | 9.72 | 9.55 | 9.45 | 9.38 | 9.29 | 9.24 | 9.20 | 9.11 | 9.03 |
| 4.06 | 4.00 | 3.94 | 3.87 | 3.83 | 3.81 | 3.77 | 3.75 | 3.74 | 3.70 | 3.67 |
| 7.87 | 7.72 | 7.56 | 7.40 | 7.30 | 7.23 | 7.14 | 7.09 | 7.06 | 6.97 | 6.89 |
| 3.64 | 3.57 | 3.51 | 3.44 | 3.40 | 3.38 | 3.34 | 3.32 | 3.30 | 3.27 | 3.23 |
| 6.62 | 6.47 | 6.31 | 6.16 | 6.06 | 5.99 | 5.91 | 5.86 | 5.82 | 5.74 | 5.66 |
| 3.35 | 3.28 | 3.22 | 3.15 | 3.11 | 3.08 | 3.04 | 3.02 | 3.01 | 2.97 | 2.93 |
| 5.81 | 5.67 | 5.52 | 5.36 | 5.26 | 5.20 | 5.12 | 5.07 | 5.03 | 4.95 | 4.87 |
| 3.14 | 3.07 | 3.01 | 2.94 | 2.89 | 2.86 | 2.83 | 2.80 | 2.79 | 2.75 | 2.71 |
| 5.26 | 5.11 | 4.96 | 4.81 | 4.71 | 4.65 | 4.57 | 4.52 | 4.48 | 4.40 | 4.32 |
| 2.98 | 2.91 | 2.85 | 2.77 | 2.73 | 2.70 | 2.66 | 2.64 | 2.62 | 2.58 | 2.54 |
| 4.85 | 4.71 | 4.56 | 4.41 | 4.31 | 4.25 | 4.17 | 4.12 | 4.08 | 4.00 | 3.92 |
| 2.85 | 2.79 | 2.72 | 2.65 | 2.60 | 2.57 | 2.53 | 2.51 | 2.49 | 2.45 | 2.41 |
| 4.54 | 4.40 | 4.25 | 4.10 | 4.01 | 3.94 | 3.86 | 3.81 | 3.78 | 3.69 | 3.61 |
| 2.75 | 2.69 | 2.62 | 2.54 | 2.50 | 2.47 | 2.43 | 2.40 | 2.38 | 2.34 | 2.30 |
| 4.30 | 4.16 | 4.01 | 3.86 | 3.76 | 3.70 | 3.62 | 3.57 | 3.54 | 3.45 | 3.37 |
| 2.67 | 2.60 | 2.53 | 2.46 | 2.41 | 2.38 | 2.34 | 2.31 | 2.30 | 2.25 | 2.21 |
| 4.10 | 3.96 | 3.82 | 3.66 | 3.57 | 3.51 | 3.43 | 3.38 | 3.34 | 3.25 | 3.18 |

TABLE F.3. Continued

| | p | \| | Degrees of freedom in the numerator | | | | | | | |
|---|---|---|---|---|---|---|---|---|---|---|
| | | 1 | 2 | 3 | 4 | 5 | 6 | 7 | 8 | 9 |
| 14 | .050 | 4.60 | 3.74 | 3.34 | 3.11 | 2.96 | 2.85 | 2.76 | 2.70 | 2.65 |
| | .010 | 8.86 | 6.51 | 5.56 | 5.04 | 4.69 | 4.46 | 4.28 | 4.14 | 4.03 |
| 15 | .050 | 4.54 | 3.68 | 3.29 | 3.06 | 2.90 | 2.79 | 2.71 | 2.64 | 2.59 |
| | .010 | 8.68 | 6.36 | 5.42 | 4.89 | 4.56 | 4.32 | 4.14 | 4.00 | 3.89 |
| 16 | .050 | 4.49 | 3.63 | 3.24 | 3.01 | 2.85 | 2.74 | 2.66 | 2.59 | 2.54 |
| | .010 | 8.53 | 6.23 | 5.29 | 4.77 | 4.44 | 4.20 | 4.03 | 3.89 | 3.78 |
| 17 | .050 | 4.45 | 3.59 | 3.20 | 2.96 | 2.81 | 2.70 | 2.61 | 2.55 | 2.49 |
| | .010 | 8.40 | 6.11 | 5.19 | 4.67 | 4.34 | 4.10 | 3.93 | 3.79 | 3.68 |
| 18 | .050 | 4.41 | 3.55 | 3.16 | 2.93 | 2.77 | 2.66 | 2.58 | 2.51 | 2.46 |
| | .010 | 8.29 | 6.01 | 5.09 | 4.58 | 4.25 | 4.01 | 3.84 | 3.71 | 3.60 |
| 19 | .050 | 4.38 | 3.52 | 3.13 | 2.90 | 2.74 | 2.63 | 2.54 | 2.48 | 2.42 |
| | .010 | 8.18 | 5.93 | 5.01 | 4.50 | 4.17 | 3.94 | 3.77 | 3.63 | 3.52 |
| 20 | .050 | 4.35 | 3.49 | 3.10 | 2.87 | 2.71 | 2.60 | 2.51 | 2.45 | 2.39 |
| | .010 | 8.10 | 5.85 | 4.94 | 4.43 | 4.10 | 3.87 | 3.70 | 3.56 | 3.46 |
| 21 | .050 | 4.32 | 3.47 | 3.07 | 2.84 | 2.68 | 2.57 | 2.49 | 2.42 | 2.37 |
| | .010 | 8.02 | 5.78 | 4.87 | 4.37 | 4.04 | 3.81 | 3.64 | 3.51 | 3.40 |
| 22 | .050 | 4.30 | 3.44 | 3.05 | 2.82 | 2.66 | 2.55 | 2.46 | 2.40 | 2.34 |
| | .010 | 7.95 | 5.72 | 4.82 | 4.31 | 3.99 | 3.76 | 3.59 | 3.45 | 3.35 |
| 23 | .050 | 4.28 | 3.42 | 3.03 | 2.80 | 2.64 | 2.53 | 2.44 | 2.37 | 2.32 |
| | .010 | 7.88 | 5.66 | 4.76 | 4.26 | 3.94 | 3.71 | 3.54 | 3.41 | 3.30 |
| 24 | .050 | 4.26 | 3.40 | 3.01 | 2.78 | 2.62 | 2.51 | 2.42 | 2.36 | 2.30 |
| | .010 | 7.82 | 5.61 | 4.72 | 4.22 | 3.90 | 3.67 | 3.50 | 3.36 | 3.26 |
| 25 | .050 | 4.24 | 3.39 | 2.99 | 2.76 | 2.60 | 2.49 | 2.40 | 2.34 | 2.28 |
| | .010 | 7.77 | 5.57 | 4.68 | 4.18 | 3.85 | 3.63 | 3.46 | 3.32 | 3.22 |
| 26 | .050 | 4.23 | 3.37 | 2.98 | 2.74 | 2.59 | 2.47 | 2.39 | 2.32 | 2.27 |
| | .010 | 7.72 | 5.53 | 4.64 | 4.14 | 3.82 | 3.59 | 3.42 | 3.29 | 3.18 |
| 27 | .050 | 4.21 | 3.35 | 2.96 | 2.73 | 2.57 | 2.46 | 2.37 | 2.31 | 2.25 |
| | .010 | 7.68 | 5.49 | 4.60 | 4.11 | 3.78 | 3.56 | 3.39 | 3.26 | 3.15 |
| 28 | .050 | 4.20 | 3.34 | 2.95 | 2.71 | 2.56 | 2.45 | 2.36 | 2.29 | 2.24 |
| | .010 | 7.64 | 5.45 | 4.57 | 4.07 | 3.75 | 3.53 | 3.36 | 3.23 | 3.12 |
| 29 | .050 | 4.18 | 3.33 | 2.93 | 2.70 | 2.55 | 2.43 | 2.35 | 2.28 | 2.22 |
| | .010 | 7.60 | 5.42 | 4.54 | 4.04 | 3.73 | 3.50 | 3.33 | 3.20 | 3.09 |
| 30 | .050 | 4.17 | 3.32 | 2.92 | 2.69 | 2.53 | 2.42 | 2.33 | 2.27 | 2.21 |
| | .010 | 7.56 | 5.39 | 4.51 | 4.02 | 3.70 | 3.47 | 3.30 | 3.17 | 3.07 |

| Degrees of freedom in the numerator | | | | | | | | | | |
|---|---|---|---|---|---|---|---|---|---|---|
| 10 | 12 | 15 | 20 | 25 | 30 | 40 | 50 | 60 | 120 | 1000 |
| 2.60 | 2.53 | 2.46 | 2.39 | 2.34 | 2.31 | 2.27 | 2.24 | 2.22 | 2.18 | 2.14 |
| 3.94 | 3.80 | 3.66 | 3.51 | 3.41 | 3.35 | 3.27 | 3.22 | 3.18 | 3.09 | 3.02 |
| 2.54 | 2.48 | 2.40 | 2.33 | 2.28 | 2.25 | 2.20 | 2.18 | 2.16 | 2.11 | 2.07 |
| 3.80 | 3.67 | 3.52 | 3.37 | 3.28 | 3.21 | 3.13 | 3.08 | 3.05 | 2.96 | 2.88 |
| 2.49 | 2.42 | 2.35 | 2.28 | 2.23 | 2.19 | 2.15 | 2.12 | 2.11 | 2.06 | 2.02 |
| 3.69 | 3.55 | 3.41 | 3.26 | 3.16 | 3.10 | 3.02 | 2.97 | 2.93 | 2.84 | 2.76 |
| 2.45 | 2.38 | 2.31 | 2.23 | 2.18 | 2.15 | 2.10 | 2.08 | 2.06 | 2.01 | 1.97 |
| 3.59 | 3.46 | 3.31 | 3.16 | 3.07 | 3.00 | 2.92 | 2.87 | 2.83 | 2.75 | 2.66 |
| 2.41 | 2.34 | 2.27 | 2.19 | 2.14 | 2.11 | 2.06 | 2.04 | 2.02 | 1.97 | 1.92 |
| 3.51 | 3.37 | 3.23 | 3.08 | 2.98 | 2.92 | 2.84 | 2.78 | 2.75 | 2.66 | 2.58 |
| 2.38 | 2.31 | 2.23 | 2.16 | 2.11 | 2.07 | 2.03 | 2.00 | 1.98 | 1.93 | 1.88 |
| 3.43 | 3.30 | 3.15 | 3.00 | 2.91 | 2.84 | 2.76 | 2.71 | 2.67 | 2.58 | 2.50 |
| 2.35 | 2.28 | 2.20 | 2.12 | 2.07 | 2.04 | 1.99 | 1.97 | 1.95 | 1.90 | 1.85 |
| 3.37 | 3.23 | 3.09 | 2.94 | 2.84 | 2.78 | 2.69 | 2.64 | 2.61 | 2.52 | 2.43 |
| 2.32 | 2.25 | 2.18 | 2.10 | 2.05 | 2.01 | 1.96 | 1.94 | 1.92 | 1.87 | 1.82 |
| 3.31 | 3.17 | 3.03 | 2.88 | 2.79 | 2.72 | 2.64 | 2.58 | 2.55 | 2.46 | 2.37 |
| 2.30 | 2.23 | 2.15 | 2.07 | 2.02 | 1.98 | 1.94 | 1.91 | 1.89 | 1.84 | 1.79 |
| 3.26 | 3.12 | 2.98 | 2.83 | 2.73 | 2.67 | 2.58 | 2.53 | 2.50 | 2.40 | 2.32 |
| 2.27 | 2.20 | 2.13 | 2.05 | 2.00 | 1.96 | 1.91 | 1.88 | 1.86 | 1.81 | 1.76 |
| 3.21 | 3.07 | 2.93 | 2.78 | 2.69 | 2.62 | 2.54 | 2.48 | 2.45 | 2.35 | 2.27 |
| 2.25 | 2.18 | 2.11 | 2.03 | 1.97 | 1.94 | 1.89 | 1.86 | 1.84 | 1.79 | 1.74 |
| 3.17 | 3.03 | 2.89 | 2.74 | 2.64 | 2.58 | 2.49 | 2.44 | 2.40 | 2.31 | 2.22 |
| 2.24 | 2.16 | 2.09 | 2.01 | 1.96 | 1.92 | 1.87 | 1.84 | 1.82 | 1.77 | 1.72 |
| 3.13 | 2.99 | 2.85 | 2.70 | 2.60 | 2.54 | 2.45 | 2.40 | 2.36 | 2.27 | 2.18 |
| 2.22 | 2.15 | 2.07 | 1.99 | 1.94 | 1.90 | 1.85 | 1.82 | 1.80 | 1.75 | 1.70 |
| 3.09 | 2.96 | 2.81 | 2.66 | 2.57 | 2.50 | 2.42 | 2.36 | 2.33 | 2.23 | 2.14 |
| 2.20 | 2.13 | 2.06 | 1.97 | 1.92 | 1.88 | 1.84 | 1.81 | 1.79 | 1.73 | 1.68 |
| 3.06 | 2.93 | 2.78 | 2.63 | 2.54 | 2.47 | 2.38 | 2.33 | 2.29 | 2.20 | 2.11 |
| 2.19 | 2.12 | 2.04 | 1.96 | 1.91 | 1.87 | 1.82 | 1.79 | 1.77 | 1.71 | 1.66 |
| 3.03 | 2.90 | 2.75 | 2.60 | 2.51 | 2.44 | 2.35 | 2.30 | 2.26 | 2.17 | 2.08 |
| 2.18 | 2.10 | 2.03 | 1.94 | 1.89 | 1.85 | 1.81 | 1.77 | 1.75 | 1.70 | 1.65 |
| 3.00 | 2.87 | 2.73 | 2.57 | 2.48 | 2.41 | 2.33 | 2.27 | 2.23 | 2.14 | 2.05 |
| 2.16 | 2.09 | 2.01 | 1.93 | 1.88 | 1.84 | 1.79 | 1.76 | 1.74 | 1.68 | 1.63 |
| 2.98 | 2.84 | 2.70 | 2.55 | 2.45 | 2.39 | 2.30 | 2.25 | 2.21 | 2.11 | 2.02 |

TABLE F.3. Continued

| | | Degrees of freedom in the numerator | | | | | | | | |
|---|---|---|---|---|---|---|---|---|---|---|
| | $p$ | 1 | 2 | 3 | 4 | 5 | 6 | 7 | 8 | 9 |
| 40 | .050 | 4.08 | 3.23 | 2.84 | 2.61 | 2.45 | 2.34 | 2.25 | 2.18 | 2.12 |
| | .010 | 7.31 | 5.18 | 4.31 | 3.83 | 3.51 | 3.29 | 3.12 | 2.99 | 2.89 |
| 50 | .050 | 4.03 | 3.18 | 2.79 | 2.56 | 2.40 | 2.29 | 2.20 | 2.13 | 2.07 |
| | .010 | 7.17 | 5.06 | 4.20 | 3.72 | 3.41 | 3.19 | 3.02 | 2.89 | 2.78 |
| 60 | .050 | 4.00 | 3.15 | 2.76 | 2.53 | 2.37 | 2.25 | 2.17 | 2.10 | 2.04 |
| | .010 | 7.08 | 4.98 | 4.13 | 3.65 | 3.34 | 3.12 | 2.95 | 2.82 | 2.72 |
| 100 | .050 | 3.94 | 3.09 | 2.70 | 2.46 | 2.31 | 2.19 | 2.10 | 2.03 | 1.97 |
| | .010 | 6.90 | 4.82 | 3.98 | 3.51 | 3.21 | 2.99 | 2.82 | 2.69 | 2.59 |
| 200 | .050 | 3.89 | 3.04 | 2.65 | 2.42 | 2.26 | 2.14 | 2.06 | 1.98 | 1.93 |
| | .010 | 6.76 | 4.71 | 3.88 | 3.41 | 3.11 | 2.89 | 2.73 | 2.60 | 2.50 |
| 1000 | .050 | 3.85 | 3.00 | 2.61 | 2.38 | 2.22 | 2.11 | 2.02 | 1.95 | 1.89 |
| | .010 | 6.66 | 4.63 | 3.80 | 3.34 | 3.04 | 2.82 | 2.66 | 2.53 | 2.43 |

| Degrees of freedom in the numerator | | | | | | | | | | |
|---|---|---|---|---|---|---|---|---|---|---|
| 10 | 12 | 15 | 20 | 25 | 30 | 40 | 50 | 60 | 120 | 1000 |
| 2.08 | 2.00 | 1.92 | 1.84 | 1.78 | 1.74 | 1.69 | 1.66 | 1.64 | 1.58 | 1.52 |
| 2.80 | 2.66 | 2.52 | 2.37 | 2.27 | 2.20 | 2.11 | 2.06 | 2.02 | 1.92 | 1.82 |
| 2.03 | 1.95 | 1.87 | 1.78 | 1.73 | 1.69 | 1.63 | 1.60 | 1.58 | 1.51 | 1.45 |
| 2.70 | 2.56 | 2.42 | 2.27 | 2.17 | 2.10 | 2.01 | 1.95 | 1.91 | 1.80 | 1.70 |
| 1.99 | 1.92 | 1.84 | 1.75 | 1.69 | 1.65 | 1.59 | 1.56 | 1.53 | 1.47 | 1.40 |
| 2.63 | 2.50 | 2.35 | 2.20 | 2.10 | 2.03 | 1.94 | 1.88 | 1.84 | 1.73 | 1.62 |
| 1.93 | 1.85 | 1.77 | 1.68 | 1.62 | 1.57 | 1.52 | 1.48 | 1.45 | 1.38 | 1.30 |
| 2.50 | 2.37 | 2.22 | 2.07 | 1.97 | 1.89 | 1.80 | 1.74 | 1.69 | 1.57 | 1.45 |
| 1.88 | 1.80 | 1.72 | 1.62 | 1.56 | 1.52 | 1.46 | 1.41 | 1.39 | 1.30 | 1.21 |
| 2.41 | 2.27 | 2.13 | 1.97 | 1.87 | 1.79 | 1.69 | 1.63 | 1.58 | 1.45 | 1.30 |
| 1.84 | 1.76 | 1.68 | 1.58 | 1.52 | 1.47 | 1.41 | 1.36 | 1.33 | 1.24 | 1.11 |
| 2.34 | 2.20 | 2.06 | 1.90 | 1.79 | 1.72 | 1.61 | 1.54 | 1.50 | 1.35 | 1.16 |

TABLE F.4. The Pearson Correlation Coefficient: $r$-Values

| df | .05 | .01 | df | .05 | .01 |
|----|-----|-----|----|-----|-----|
| 1 | .997 | 1.000 | 31 | .344 | .442 |
| 2 | .950 | .990 | 32 | .339 | .436 |
| 3 | .878 | .959 | 33 | .334 | .430 |
| 4 | .812 | .917 | 34 | .329 | .424 |
| 5 | .755 | .875 | 35 | .325 | .418 |
| 6 | .707 | .834 | 36 | .320 | .413 |
| 7 | .666 | .798 | 37 | .316 | .408 |
| 8 | .632 | .765 | 38 | .312 | .403 |
| 9 | .602 | .735 | 39 | .308 | .398 |
| 10 | .576 | .708 | 40 | .304 | .393 |
| 11 | .553 | .684 | 41 | .301 | .389 |
| 12 | .533 | .661 | 42 | .297 | .384 |
| 13 | .514 | .641 | 43 | .294 | .380 |
| 14 | .497 | .623 | 44 | .291 | .376 |
| 15 | .482 | .606 | 45 | .288 | .372 |
| 16 | .468 | .590 | 46 | .285 | .368 |
| 17 | .456 | .575 | 47 | .282 | .365 |
| 18 | .444 | .562 | 48 | .279 | .361 |
| 19 | .433 | .549 | 49 | .276 | .358 |
| 20 | .423 | .537 | 50 | .273 | .354 |
| 21 | .413 | .526 | 60 | .250 | .325 |
| 22 | .404 | .515 | 70 | .232 | .302 |
| 23 | .396 | .505 | 80 | .217 | .283 |
| 24 | .388 | .496 | 90 | .205 | .267 |
| 25 | .381 | .487 | 100 | .195 | .254 |
| 26 | .374 | .479 | 200 | .138 | .181 |
| 27 | .367 | .471 | 300 | .113 | .148 |
| 28 | .361 | .463 | 400 | .098 | .128 |
| 29 | .355 | .456 | 500 | .088 | .115 |
| 30 | .349 | .449 | 1,000 | .062 | .081 |

TABLE F.5. Critical $r_s$ Values for the Spearman Correlation Coefficient

| | Level of Significance ($\alpha$) for One-Tailed Test | | | | | | | | |
| | .25 | .10 | .05 | .025 | .01 | .005 | .0025 | .001 | .0005 |
| | Level of Significance ($\alpha$) for Two-Tailed Test | | | | | | | | |
| n | .50 | .20 | .10 | .05 | .02 | .01 | .005 | .002 | .001 |
|---|---|---|---|---|---|---|---|---|---|
| 4 | .600 | 1.000 | 1.000 | | | | | | |
| 5 | .500 | .800 | .900 | 1.000 | 1.000 | | | | |
| 6 | .371 | .657 | .829 | .886 | .943 | 1.000 | 1.000 | | |
| 7 | .321 | .571 | .714 | .786 | .893 | .929 | .964 | 1.000 | 1.000 |
| 8 | .310 | .524 | .643 | .738 | .833 | .881 | .905 | .952 | .976 |
| 9 | .267 | .483 | .600 | .700 | .783 | .833 | .867 | .917 | .933 |
| 10 | .248 | .455 | .564 | .648 | .745 | .794 | .830 | .879 | .903 |
| 11 | .236 | .427 | .536 | .618 | .709 | .755 | .800 | .845 | .873 |
| 12 | .217 | .406 | .503 | .587 | .678 | .727 | .769 | .818 | .846 |
| 13 | .209 | .385 | .484 | .560 | .648 | .703 | .747 | .791 | .824 |
| 14 | .200 | .367 | .464 | .538 | .626 | .679 | .723 | .771 | .802 |
| 15 | .189 | .354 | .446 | .521 | .604 | .654 | .700 | .750 | .779 |
| 16 | .182 | .341 | .429 | .503 | .582 | .635 | .679 | .729 | .762 |
| 17 | .176 | .328 | .414 | .485 | .566 | .615 | .662 | .713 | .748 |
| 18 | .170 | .317 | .401 | .472 | .550 | .600 | .643 | .695 | .728 |
| 19 | .165 | .309 | .391 | .460 | .535 | .584 | .628 | .677 | .712 |
| 20 | .161 | .299 | .380 | .447 | .520 | .570 | .612 | .662 | .696 |
| 21 | .156 | .292 | .370 | .435 | .508 | .556 | .599 | .648 | .681 |
| 22 | .152 | .284 | .361 | .425 | .496 | .544 | .586 | .634 | .667 |
| 23 | .148 | .278 | .353 | .415 | .486 | .532 | .573 | .622 | .654 |
| 24 | .144 | .271 | .344 | .406 | .476 | .521 | .562 | .610 | .642 |
| 25 | .142 | .265 | .337 | .398 | .466 | .511 | .551 | .598 | .630 |
| 26 | .138 | .259 | .331 | .390 | .457 | .501 | .541 | .587 | .619 |
| 27 | .136 | .255 | .324 | .382 | .448 | .491 | .531 | .577 | .608 |
| 28 | .133 | .250 | .317 | .375 | .440 | .483 | .522 | .567 | .598 |
| 29 | .130 | .245 | .312 | .368 | .433 | .475 | .513 | .558 | .589 |
| 30 | .128 | .240 | .306 | .362 | .425 | .467 | .504 | .549 | .580 |
| 31 | .126 | .236 | .301 | .356 | .418 | .459 | .496 | .541 | .571 |
| 32 | .124 | .232 | .296 | .350 | .412 | .452 | .489 | .533 | .563 |
| 33 | .121 | .229 | .291 | .345 | .405 | .446 | .482 | .525 | .554 |
| 34 | .120 | .225 | .287 | .340 | .399 | .439 | .475 | .517 | .547 |
| 35 | .118 | .222 | .283 | .335 | .394 | .433 | .468 | .510 | .539 |

TABLE F.5. Continued

| | Level of Significance (α) for One-Tailed Test | | | | | | | | |
|---|---|---|---|---|---|---|---|---|---|
| | .25 | .10 | .05 | .025 | .01 | .005 | .0025 | .001 | .0005 |
| | Level of Significance (α) for Two-Tailed Test | | | | | | | | |
| n | .50 | .20 | .10 | .05 | .02 | .01 | .005 | .002 | .001 |
| 36 | .116 | .219 | .279 | .330 | .388 | .427 | .462 | .504 | .533 |
| 37 | .114 | .216 | .275 | .325 | .383 | .421 | .456 | .497 | .526 |
| 38 | .113 | .212 | .271 | .321 | .378 | .415 | .450 | .491 | .519 |
| 39 | .111 | .210 | .267 | .317 | .373 | .410 | .444 | .485 | .513 |
| 40 | .110 | .207 | .264 | .313 | .368 | .405 | .439 | .479 | .507 |
| 41 | .108 | .204 | .261 | .309 | .364 | .400 | .433 | .473 | .501 |
| 42 | .107 | .202 | .257 | .305 | .359 | .395 | .428 | .468 | .495 |
| 43 | .105 | .199 | .254 | .301 | .355 | .391 | .423 | .463 | .490 |
| 44 | .104 | .197 | .251 | ..298 | .351 | .386 | .419 | .458 | .484 |
| 45 | .103 | .194 | .248 | .294 | .347 | .382 | .414 | .453 | .479 |
| 46 | .102 | .192 | .246 | .291 | .343 | .378 | .410 | .448 | .474 |
| 47 | .101 | .190 | .243 | .288 | .340 | .374 | .405 | .443 | .469 |
| 48 | .100 | .188 | .240 | .285 | .336 | .370 | .401 | .439 | .465 |
| 49 | .098 | .186 | .238 | .282 | .333 | .366 | .397 | .434 | .460 |
| 50 | .097 | .184 | .235 | .279 | .329 | .363 | .393 | .430 | .456 |
| 51 | .096 | .182 | .233 | .276 | .326 | .359 | .390 | .426 | .451 |
| 52 | .095 | .180 | .231 | .274 | .323 | .356 | .386 | .422 | .447 |
| 53 | .095 | .179 | .228 | .271 | .320 | .352 | .382 | .418 | .443 |
| 54 | .094 | .177 | .226 | .268 | .317 | .349 | .379 | .414 | .439 |
| 55 | .093 | .175 | .224 | .266 | .314 | .346 | .375 | .411 | .435 |
| 56 | .092 | .174 | .222 | .264 | .311 | .343 | .372 | .407 | .432 |
| 57 | .091 | .172 | .220 | .261 | .308 | .340 | .369 | .404 | .428 |
| 58 | .090 | .171 | .218 | .259 | .306 | .337 | .366 | .400 | .424 |
| 59 | .089 | .169 | .216 | .257 | .303 | .334 | .363 | .397 | .421 |
| 60 | .089 | .168 | .214 | .255 | .300 | .331 | .360 | .394 | .418 |
| 61 | .088 | .166 | .213 | .252 | .298 | .329 | .357 | .391 | .414 |
| 62 | .087 | .165 | .211 | .250 | .296 | .326 | .354 | .388 | .411 |
| 63 | .086 | .163 | .209 | .248 | .293 | .323 | .351 | .385 | .408 |
| 64 | .086 | .162 | .207 | .246 | .291 | .321 | .348 | .382 | .405 |
| 65 | .085 | .161 | .206 | .244 | .289 | .318 | .346 | .379 | .402 |
| 66 | .084 | .160 | .204 | .243 | .287 | .316 | .343 | .376 | .399 |
| 67 | .084 | .158 | .203 | .241 | .284 | .314 | .341 | .373 | .396 |
| 68 | .083 | .157 | .201 | .239 | .282 | .311 | .338 | .370 | .393 |

TABLE F.5. Continued

| | Level of Significance ($\alpha$) for One-Tailed Test | | | | | | | | |
|---|---|---|---|---|---|---|---|---|---|
| | .25 | .10 | .05 | .025 | .01 | .005 | .0025 | .001 | .0005 |
| | Level of Significance ($\alpha$) for Two-Tailed Test | | | | | | | | |
| n | .50 | .20 | .10 | .05 | .02 | .01 | .005 | .002 | .001 |
| 69 | .082 | .156 | .200 | .237 | .280 | .309 | .336 | .368 | .390 |
| 70 | .082 | .155 | .198 | .235 | .278 | .307 | .333 | .365 | .388 |
| 71 | .081 | .154 | .197 | .234 | .276 | .305 | .331 | .363 | .385 |
| 72 | .081 | .153 | .195 | .232 | .274 | .303 | .329 | .360 | .382 |
| 73 | .080 | .152 | .194 | .230 | .272 | .301 | .327 | .358 | .380 |
| 74 | .080 | .151 | .193 | .229 | .271 | .299 | .324 | .355 | .377 |
| 75 | .079 | .150 | .191 | .227 | .269 | .297 | .322 | .353 | .375 |
| 76 | .078 | .149 | .190 | .226 | .267 | .295 | .320 | .351 | .372 |
| 77 | .078 | .148 | .189 | .224 | .265 | .293 | .318 | .349 | .370 |
| 78 | .077 | .147 | .188 | .223 | .264 | .291 | .316 | .346 | .368 |
| 79 | .077 | .146 | .186 | .221 | .262 | .289 | .314 | .344 | .365 |
| 80 | .076 | .145 | .185 | .220 | .260 | .287 | .312 | .342 | .363 |
| 81 | .076 | .144 | .184 | .219 | .259 | .285 | .310 | .340 | .361 |
| 82 | .075 | .143 | .183 | .217 | .257 | .284 | .308 | .338 | .359 |
| 83 | .075 | .142 | .182 | .216 | .255 | .282 | .306 | .336 | .357 |
| 84 | .074 | .141 | .181 | .215 | .254 | .280 | .305 | .334 | .355 |
| 85 | .074 | .140 | .180 | .213 | .252 | .279 | .303 | .332 | .353 |
| 86 | .074 | .139 | .179 | .212 | .251 | .277 | .301 | .330 | .351 |
| 87 | .073 | .139 | .177 | .211 | .250 | .276 | .299 | .328 | .349 |
| 88 | .073 | .138 | .176 | .210 | .248 | .274 | .298 | .327 | .347 |
| 89 | .072 | .137 | .175 | .209 | .247 | .272 | .296 | .325 | .345 |
| 90 | .072 | .136 | .174 | .207 | .245 | .271 | .294 | .323 | .343 |
| 91 | .072 | .135 | .173 | .206 | .244 | .269 | .293 | .321 | .341 |
| 92 | .071 | .135 | .173 | .205 | .243 | .268 | .291 | .319 | .339 |
| 93 | .071 | .134 | .172 | .204 | .241 | .267 | .290 | .318 | .338 |
| 94 | .070 | .133 | .171 | .203 | .240 | .265 | .288 | .316 | .336 |
| 95 | .070 | .133 | .170 | .202 | .239 | .264 | .287 | .314 | .334 |
| 96 | .070 | .132 | .169 | .201 | .238 | .262 | .285 | .313 | .332 |
| 97 | .069 | .131 | .168 | .200 | .236 | .261 | .284 | .311 | .331 |
| 98 | .069 | .130 | .167 | .199 | .235 | .260 | .282 | .310 | .329 |
| 99 | .068 | .130 | .166 | ..198 | .234 | .258 | .281 | .308 | .327 |
| 100 | .068 | .129 | .165 | .197 | .233 | .257 | .279 | .307 | .326 |

TABLE F.6. Critical Values of Chi Square

| df | **Area to the Right of the Critical Value** | | | | |
|----|------|------|------|------|------|
|    | .1 | .05 | .025 | .01 | .001 |
| 1  | 2.706 | 3.841 | 5.024 | 6.635 | 10.828 |
| 2  | 4.605 | 5.991 | 7.378 | 9.210 | 13.816 |
| 3  | 6.251 | 7.815 | 9.348 | 11.345 | 16.266 |
| 4  | 7.779 | 9.488 | 11.143 | 13.277 | 18.467 |
| 5  | 9.236 | 11.070 | 12.833 | 15.086 | 20.515 |
| 6  | 10.645 | 12.592 | 14.449 | 16.812 | 22.458 |
| 7  | 12.017 | 14.067 | 16.013 | 18.475 | 24.322 |
| 8  | 13.362 | 15.507 | 17.535 | 20.090 | 26.125 |
| 9  | 14.684 | 16.919 | 19.023 | 21.666 | 27.877 |
| 10 | 15.987 | 18.307 | 20.483 | 23.209 | 29.588 |
| 11 | 17.275 | 19.675 | 21.920 | 24.725 | 31.264 |
| 12 | 18.549 | 21.026 | 23.337 | 26.217 | 32.910 |
| 13 | 19.812 | 22.362 | 24.736 | 27.688 | 34.528 |
| 14 | 21.064 | 23.685 | 26.119 | 29.141 | 36.123 |
| 15 | 22.307 | 24.996 | 27.488 | 30.578 | 37.697 |
| 16 | 23.542 | 26.296 | 28.845 | 32.000 | 39.252 |
| 17 | 24.769 | 27.587 | 30.191 | 33.409 | 40.790 |
| 18 | 25.989 | 20.069 | 31.526 | 34.805 | 42.312 |
| 19 | 27.204 | 30.144 | 32.852 | 36.191 | 43.820 |
| 20 | 28.412 | 31.410 | 34.170 | 37.566 | 45.315 |
| 21 | 29.615 | 32.671 | 35.479 | 38.932 | 46.797 |
| 22 | 30.813 | 33.924 | 36.781 | 40.289 | 48.268 |
| 23 | 32.007 | 35.172 | 38.076 | 41.638 | 49.728 |
| 24 | 33.196 | 36.415 | 39.364 | 42.980 | 51.179 |
| 25 | 34.382 | 37.652 | 40.646 | 44.314 | 52.620 |
| 26 | 35.563 | 38.885 | 41.923 | 45.642 | 54.052 |
| 27 | 36.741 | 40.113 | 43.195 | 46.963 | 55.476 |
| 28 | 37.916 | 41.337 | 44.461 | 48.278 | 56.892 |
| 29 | 39.087 | 42.557 | 45.722 | 49.588 | 58.301 |
| 30 | 40.256 | 43.773 | 46.979 | 50.892 | 59.703 |
| 31 | 41.422 | 44.985 | 48.232 | 52.191 | 61.098 |
| 32 | 42.585 | 46.194 | 49.480 | 53.486 | 62.487 |
| 33 | 43.745 | 47.400 | 50.725 | 54.776 | 63.870 |
| 34 | 44.903 | 48.602 | 51.966 | 56.061 | 65.247 |

TABLE F.6. Continued

### Area to the Right of the Critical Value

| df | .1 | .05 | .025 | .01 | .001 |
|----|------|------|------|------|------|
| 35 | 46.059 | 49.802 | 53.203 | 57.342 | 66.619 |
| 36 | 47.212 | 50.998 | 54.437 | 58.619 | 67.985 |
| 37 | 48.363 | 52.192 | 55.668 | 59.893 | 69.347 |
| 38 | 49.513 | 53.384 | 56.896 | 61.162 | 70.703 |
| 39 | 50.660 | 54.572 | 58.120 | 62.428 | 72.055 |
| 40 | 51.805 | 55.758 | 59.342 | 63.691 | 73.402 |
| 41 | 52.949 | 56.942 | 60.561 | 64.950 | 74.745 |
| 42 | 54.090 | 58.124 | 61.777 | 66.206 | 76.084 |
| 43 | 55.230 | 59.304 | 62.990 | 67.459 | 77.419 |
| 44 | 56.369 | 60.481 | 64.201 | 68.710 | 78.750 |
| 45 | 57.505 | 61.656 | 65.410 | 69.957 | 80.077 |
| 46 | 58.641 | 62.830 | 66.617 | 71.201 | 81.400 |
| 47 | 59.774 | 64.001 | 67.821 | 72.443 | 82.720 |
| 48 | 60.907 | 65.171 | 69.023 | 73.683 | 84.037 |
| 49 | 62.038 | 66.339 | 70.222 | 74.919 | 85.351 |
| 50 | 63.167 | 67.505 | 71.420 | 76.154 | 86.661 |
| 51 | 64.295 | 68.669 | 72.616 | 77.386 | 87.968 |
| 52 | 65.422 | 69.832 | 73.810 | 78.616 | 89.272 |
| 53 | 66.548 | 70.993 | 75.002 | 79.843 | 90.573 |
| 54 | 67.673 | 72.153 | 76.192 | 81.069 | 91.872 |
| 55 | 68.796 | 73.311 | 77.380 | 82.292 | 93.168 |
| 56 | 69.919 | 74.468 | 78.567 | 83.513 | 94.461 |
| 57 | 71.040 | 75.624 | 79.752 | 84.733 | 95.751 |
| 58 | 72.160 | 76.778 | 80.936 | 85.950 | 97.039 |
| 59 | 73.279 | 77.931 | 82.117 | 87.166 | 98.324 |
| 60 | 74.397 | 79.082 | 83.298 | 88.379 | 99.607 |
| 61 | 75.514 | 80.232 | 84.476 | 89.591 | 100.888 |
| 62 | 76.630 | 81.381 | 85.654 | 90.802 | 102.166 |
| 63 | 77.745 | 82.529 | 86.830 | 92.010 | 103.442 |
| 64 | 78.860 | 83.675 | 88.004 | 93.217 | 104.716 |
| 65 | 79.973 | 84.821 | 89.177 | 94.422 | 105.988 |
| 66 | 81.085 | 85.965 | 90.349 | 95.626 | 107.258 |
| 67 | 82.197 | 87.108 | 91.519 | 96.828 | 108.526 |

TABLE F.6. Continued

**Area to the Right of the Critical Value**

| df | .1 | .05 | .025 | .01 | .001 |
|----|------|------|------|------|------|
| 68 | 83.308 | 88.250 | 92.689 | 98.028 | 109.791 |
| 69 | 84.418 | 89.391 | 93.856 | 99.228 | 111.055 |
| 70 | 85.527 | 90.531 | 95.023 | 100.425 | 112.317 |
| 71 | 86.635 | 91.670 | 96.189 | 101.621 | 113.577 |
| 72 | 87.743 | 92.808 | 97.353 | 102.816 | 114.835 |
| 73 | 88.850 | 93.945 | 98.516 | 104.010 | 116.092 |
| 74 | 89.956 | 95.081 | 99.678 | 105.202 | 117.346 |
| 75 | 91.061 | 96.217 | 100.839 | 106.393 | 118.599 |
| 76 | 92.166 | 97.351 | 101.999 | 107.583 | 119.850 |
| 77 | 93.270 | 98.484 | 103.158 | 108.771 | 121.100 |
| 78 | 94.374 | 99.617 | 104.316 | 109.958 | 122.348 |
| 79 | 95.476 | 100.749 | 105.473 | 111.144 | 123.594 |
| 80 | 96.578 | 101.879 | 106.629 | 112.329 | 124.839 |
| 81 | 97.680 | 103.010 | 107.783 | 113.512 | 126.083 |
| 82 | 98.780 | 104.139 | 108.937 | 114.695 | 127.324 |
| 83 | 99.880 | 105.267 | 110.090 | 115.876 | 128.565 |
| 84 | 100.980 | 106.395 | 111.242 | 117.057 | 129.804 |
| 85 | 102.079 | 107.522 | 112.393 | 118.236 | 131.041 |
| 86 | 103.177 | 108.648 | 113.544 | 119.414 | 132.277 |
| 87 | 104.275 | 109.773 | 114.693 | 120.591 | 133.512 |
| 88 | 105.372 | 110.898 | 115.841 | 121.767 | 134.746 |
| 89 | 106.469 | 112.022 | 116.989 | 122.942 | 135.978 |
| 90 | 107.565 | 113.145 | 118.136 | 124.116 | 137.208 |
| 91 | 108.661 | 114.268 | 119.282 | 125.289 | 138.438 |
| 92 | 109.756 | 115.390 | 120.427 | 126.462 | 139.666 |
| 93 | 110.850 | 116.511 | 121.571 | 127.633 | 140.893 |
| 94 | 111.944 | 117.632 | 122.715 | 128.803 | 142.119 |
| 95 | 113.038 | 118.752 | 123.858 | 129.973 | 143.344 |
| 96 | 114.131 | 119.871 | 125.000 | 131.141 | 144.567 |
| 97 | 115.223 | 120.990 | 126.141 | 132.309 | 145.789 |
| 98 | 116.315 | 122.108 | 127.282 | 133.476 | 147.010 |
| 99 | 117.407 | 123.225 | 128.422 | 134.642 | 148.230 |
| 100 | 118.498 | 124.342 | 129.561 | 135.807 | 149.449 |

# REFERENCES

Allen, A. O. (1990). *Probability, statistics, and queueing theory: With computer science applications* (2nd ed.). San Diego, CA: Academic Press. Retrieved from http://books.google.com/books?hl=en&lr=&id=PMMUbHvr-7sC&oi=fnd&pg=PR11&dq=arnold,+1990+%2B+statistics&ots=ANCEXzLEBV&sig=42rSmNpMCJm0b3e04gsF3ZZKEIQ

Auerbach, C., & Schudrich, W. Z. (2013). SSD for R: A comprehensive statistical package to analyze single-system data. *Research on Social Work Practice, 23*(3), 346–353.

Auerbach, C., & Zeitlin, W. (2014). *SSD for R: An R package for analyzing single-subject data.* New York: Oxford University Press.

Auerbach, C., & Zeitlin, W. (2015). *Making your case: Using R for program evaluation.* New York: Oxford University Press.

Bloom, M., Fischer, J., & Orme, J. G. (2009). *Evaluating practice: Guidelines for the accountable professional* (6th ed.). New York, NY: Pearson.

Casella, G., & Berger, R. L. (1990). *Statistical inference* (Vol. 70). Belmont, CA: Duxbury Press. Retrieved from http://departments.columbian.gwu.edu/statistics/sites/default/files/u20/Syllabus%206202-Spring%202013-%20Li.pdf

Cherry, S. (1998). Statistical tests in publications of The Wildlife Society. *Wildlife Society Bulletin, 26*(4), 947–953.

Cohen, J. (1988). *Statistical power analysis for the behavioral sciences* (2nd ed.). Hillsdale, NJ: Lawrence Erlbaum Associates.

Crawley, M. J. (2012). *The R book.* John Wiley & Sons.

Fox, J., & Weisberg, S. (2011). *An R companion to applied regression* (2nd ed.). Thousand Oaks, CA: Sage Publications.

Franco, A., Malhotra, N., & Simonovits, G. (2014). Publication bias in the social sciences: Unlocking the file drawer. *Science, 345*(6203), 1502–1505.

Freedman, D., & Diaconis, P. (1981). On the histogram as a density estimator: L 2 theory. *Zeitschrift Für Wahrscheinlichkeitstheorie Und Verwandte Gebiete, 57*(4), 453–476. Retrieved from https://doi.org/10.1007/BF01025868

Gandrud, C. (2015). *Reproducible research with R and R studio* (2nd ed.). New York, NY: CRC Press. Retrieved from https://englianhu.files.wordpress.com/2016/01/reproducible-research-with-r-and-studio-2nd-edition.pdf

Kabacoff, R. (2011). *R in action: Data analysis and graphics with R.* Shelter Island, NY: Manning.

Moore, D. S., & McCabe, G. P. (1989). *Introduction to the practice of statistics.* New York, NY: W. H. Freeman.

Murphy, K. R., Myors, B., & Wolach, A. (2014). *Statistical power analysis: A simple and general model for traditional and modern hypothesis tests* (4th ed.). New York, NY: Routledge.

The R Project for Statistical Computing. (n.d.). *What is R?* Retrieved from http://www.r-project.org/about.html.Schneider, J. W. (2013). Caveats for using statistical significance tests in research assessments. *Journal of Informetrics, 7*(1), 50–62.

Verzani, J. (2004). *Using R for introductory statistics.* Boca Raton, FL: Chapman and Hall/CRC.

Welch, B. L. (1938). The significance of the difference between two means when the population variances are unequal. *Biometrika, 29*(3/4), 350–362.

# INDEX

Tables and figures are indicated by an italic *t* and *f* following the paragraph number